SPACE ODYSSEY

SPACE ODYSSEY

A JOURNEY TO THE PLANETS

TIM HAINES AND
CHRISTOPHER RILEY

This book is published to accompany the BBC Television series
Space Odyssey first broadcast on BBC 1 in 2004

Executive producers: Tim Haines and Adam Kemp
Series producer: Christopher Riley

Published by BBC Books,
BBC Worldwide Ltd
Woodlands
80 Wood Lane
London W12 OTT

First published 2004
Text © Tim Haines and Christopher Riley 2004
The moral right of the authors has been asserted.

ISBN 0 563 52154 6

Commissioning Editor: *Shirley Patton*
Project Editor: *Helena Caldon*
Copy Editor: *Ben Morgan*
Designer: *Bobby Birchall* at *DW Design*
Cover Art Director: *Pene Parker*
Picture Researchers: *Duncan Copp* and *Miriam Hyman*
Production Controller: *Arlene Alexander*

Colour origination and printing, in Great Britain by Butler & Tanner

For more information about this and other BBC books, please visit
our website on www.bbcshop.com

1. Launch of *Vostok 1* carrying Yuri Gagarin, the first man to go
   into space (1961).
2. Cosmonauts Pavel Belyayev and Alexei Leonov in the cabin of
   Voskhod-2 (1965).
3a&3b. President Nixon congratulates astronauts Neil Armstrong and
   Buzz Aldrin on the success of the first lunar landing (1969).
4. *Apollo 11* launch (1969).
5. Astronaut Eugene Cernan, the last man on the Moon, on the
   lunar rover (1972).
6. *Skylab* in Earth orbit (1973).
7. Cosmonaut Svetlana Savitskay – the first woman to walk in
   space (1984).
8. The Russian *Mir* space station seen from *Space Shuttle
   Atlantis* STS-79 (1996).
9. The first two modules of the *International Space Station*: *Zarya*
   and *Unity* in orbit (1998).

PREFACE

For over a generation no human has ventured beyond Earth orbit. Recent calls for a return to the Moon and then a journey on to Mars are still decades away if they happen at all. The pace of human space exploration has slowed and the tragedies of *Columbia* and *Challenger* have thrown the role of humans in space into doubt. Despite this, we are in fact living in a great age of planetary exploration. It's just that our increasingly autonomous robots are doing the exploring for us. In the absence of human flights to the planets it is often forgotten that robotic probes have been immensely successful over the last few decades.

Thanks to the 'magic' of robotic space exploration, we have been able to tell our story with more science and less fiction. The details of our human experiences on these other worlds are based on the facts relayed to us from our robotic missions. No laws of physics are broken and, although the propulsion systems and magnetic shields we have harnessed to make the mission possible only exist in theory, they are more imaginable to us today than Concorde would have been to the Wright brothers. The one area our probes cannot help us with is the psychology of long-duration space flights. Although Russian cosmonauts have pushed the endurance records to 747 cumulative days this is still a long way short of the 2240 days we have calculated it would take *Pegasus* to

Armed with the knowledge from 40 years of this space exploration and unencumbered by the reality of budgets and bureaucracy, we have imagined our way around the solar system. If these stories in the following pages sound fanciful, perhaps even like magic, think what Jules Verne's story *Voyage Dans la Lune* about a voyage to the Moon penned in 1902 must have sounded like when it was published 67 years before the Apollo landings. Or what Christian Huygens, the seventeenth-century clockmaker and astronomer, would have made of the robotic lander named after him bound for Saturn's moon, Titan. The dizzy speed of technological progress is succinctly summed up by Arthur C. Clarke's third law: 'Sufficiently advanced technology is indistinguishable from magic.'

complete our Grand Tour. From conversations with many astronauts it is clear they pride themselves on being the sort of people that will endure far more physically and emotionally than most of us. So we have assumed that despite the monotony, the danger and the claustrophobic conditions none of our crew 'cracks'. Instead they are driven by a sense of duty, comradeship and, in the end, sheer bloody-mindedness to complete the journey.

A generation has grown up not knowing a time when humans could fly to other worlds. Until we decide to go once more, this book is for them.

Tim Haines and Christopher Riley, summer 2004

'Earth is the cradle of humanity, but one

cannot remain in the cradle forever.'

Konstantin Tsiolkovsky 1911

INTRODUCTION

On 6 September 1522 a battered and distressed ship nosed its way stubbornly into the harbour of Sanlucar de Barrameda in southern Spain. Her main mast was close to breaking off and she leaked so badly that her crew were pumping round the clock to keep her afloat. Disease, starvation and severe storms had reduced the crew of 50 to just 18 half-dead men. That the *Victoria* had survived at all was a glorious thing. Three years before, when she left the safe confines of coastal waters and set off across the open ocean, many never expected to see her again; yet she sailed all the same. And so she became the first ship to sail round the world.

Such giant leaps of progress define our species. We are driven by the urge to explore and discover, and within the last century the speed of progress has accelerated. Less than 450 years after the *Victoria*'s voyage, the Wright brothers completed the first powered flight. Only 58 years later, Yuri Gagarin became the first person to fly in space, circumnavigating the world in just 90 minutes. Eight years after that, Neil Armstrong and Buzz Aldrin had walked on another world.

The voyage of *Pegasus* would carry us further to other worlds.

The whole endeavour began at the dawn of a new century and against a background of understandable scepticism from some quarters. It was proposed that a manned spacecraft should attempt a grand tour – a voyage around the solar system. Plans for nuclear-powered rockets had been on the drawing board since the 1960s, and successful robotic missions had repeatedly proved our ability to navigate through space and send craft to the planets with pinpoint accuracy. Even the long-term endurance of zero gravity by humans had been pushed to over two years. And of course there had been an exponential growth in the power of computers.

Put in these terms, the project was not as impossible as it first sounded. Indeed, we had a better picture of where we were heading than Christopher Columbus had when his ship first slipped over the Atlantic horizon and into the unknown, en route for the Americas. Almost 50 years of robotic exploration of the planets had paved the way for humans to follow. Our Pioneers, Mariners, Veneras, Vikings, Voyagers and Pathfinders had imaged, charted, mapped and scrutinized our neighbouring worlds to show us the way. So, the critics argued, why send humans? Let the robots continue their fine work – it would cost a lot less money. The counter-argument was that robots had done their job already; it was time to move on.

Drawing on the experiences of the Apollo and Zond manned lunar missions, and on the results of space-habitation studies carried out on *Skylab*, the Salyuts, *Mir*, *Shenzhou* and the *International Space Station*, a new breed of spacecraft began to take shape in Earth orbit. This was *Pegasus*. At 1.3 km long and weighing over 400 tonnes, it was one of the largest metal structures ever constructed – and certainly the largest structure ever built in space. In the tradition of Victorian exploration, fuel and food dumps were placed strategically across the solar system. Constellations of communication satellites were dispatched ahead of *Pegasus* to the destination planets, and armadas of other robots were positioned across the inner solar system to watch for solar storms that might threaten our vulnerable crew outside the protective shield of Earth's atmosphere and magnetosphere.

The positions and orbits of the planets had sealed the route of our mission billions of years before. *Pegasus* would visit Venus, Mars, Jupiter, Saturn and Pluto, before returning safely to Earth in just over six years.

The principal scientific goal was to search for signs of life – or at least evidence of past life. From the cloud tops of Venus to the perhaps still watery subsurface of Mars, from the hot, sulphurous vents of Io to the oceans of Europa, and from the rings of Saturn to the icy hollows of a comet, the crew of *Pegasus* would scour the solar system. Beyond this keystone objective, the mission aimed to map landscapes and sample atmospheres, adding detail to our existing knowledge and so helping to shine light on the origin and evolution of the 100-odd planets, moons and countless minor worlds that make up the solar system.

Ships anchored at Cape Kennedy in sight of a spaceship bound for a new sea.

Such extraordinary objectives could only be realized under extraordinary circumstances. The challenge of *Pegasus* was met by the exceptional dedication and skill of millions of engineers, scientists and technicians across the world; their selfless commitment ensured the project's success. Most will remain nameless, but five individuals will be remembered by history – commander Tom Kirby, astronaut medic John Pearson, flight engineer Yvan Grigorev, and mission scientists Nina Sulman and Zoë Lessard.

As chief scientist for Project Pegasus, I was thankful to be alive at the moment in history when humans first left the Earth–Moon system. I watched my five friends explore new worlds, sometimes wishing, perhaps like many of you, that I was among them. In this book we tell the story of *Pegasus* in the words of the people who flew her, written as these momentous events took place. I hope that future generations will learn from these pages what it felt like to leave our cradle, the Earth, and take giant strides across space to walk on other worlds.

Alex Lloyd
Chief Scientist, *Pegasus* Mission

Pathway to the Planets

No tour of the solar system is simply a matter of powering directly from planet to planet. The dates of a mission must be chosen wisely to take advantage of favourable planetary positions, harnessing the planet's orbits of the Sun to haul us across the void.

Pegasus left Earth orbit, where it was constructed, on a trajectory opposite to the orbit of the planets. During the first leg of the mission it fell towards the Sun to be caught by Venus's gravity 41 days after departure. The craft then spent 14 days in orbit around Venus and sent down a two-person lander called *Orpheus* for a short 'walk' on the planet's blistering surface.

Leaving Venusian orbit, *Pegasus* fired up its rockets and set off for Mars. The crossing of 241 million km took 62 days, after which the craft slowed down by firing its rockets in reverse ('retro-burning') and dipping into the planet's upper atmosphere ('aerobraking'). After slowing sufficiently, *Pegasus* became caught in orbit around Mars. It docked with a supply craft carrying fuel and food, and released a three-person lander called *Ares*, which spent 15 days on the planet's surface.

From Mars, *Pegasus* returned to the inner solar system, falling towards the Sun to pick up speed. The phenomenal acceleration this generated helped save mission time by propelling *Pegasus* 1060 million km to Jupiter in just over 270 days. Although the Sun's gravity slowed *Pegasus* during the outward journey to Jupiter, the craft still had to perform a severe aerobraking manoeuvre at Jupiter in order to place it on a course to the moon Io. A descent to the surface of Io was attempted using a single-person lander called *Hermes*; an unmanned probe was dropped into Jupiter's atmosphere; and a separate mission visited the moon Europa to collect samples from the surface. Human exploration of the Jovian system was limited to 30 days due to Jupiter's intense radiation field.

A sustained engine burn accelerated *Pegasus* to more than 290,000 km/h to make the flight to Saturn, 2 billion km away, in around 300 days. A series of aerobraking manoeuvres then placed *Pegasus* in an elliptical orbit around Saturn, and a delicate series of engine burns trimmed this orbit to place the craft in a wide gap within the rings – the Cassini Division. A daring spacewalk was conducted to collect samples of the ring fragments. Before leaving Saturn *Pegasus* headed for Titan, its largest moon, where an unmanned exploration lander was dropped through the clouds to the surface.

The crew could have returned to Earth at this point but they decided to push on, taking *Pegasus* on an audacious journey to the edge of the solar system for an encounter with Pluto. The craft used Saturn's gravity to fling it upwards and out of the plane of the solar system, and a series of

Opposite: The crew. Back row left to right – Flight Engineer Yvan Grigorev, Commander Tom Kirby, Flight Medic John Pearson.
Front row – Mission Scientists Nina Sulman and Zoë Lessard.
Above: *Orpheus* – The Venus lander.
Below: *Ares* – The Mars lander.
Previous pages: *Pegasus* departs from the spectacular comet Jano–Moore – its two brilliant tails growing in size as it approaches the Sun.

lengthy engine burns accelerated it to 80 km a second. Even at this speed, the 4.2 billion km to Pluto took more than 600 days. Catching the tiny planet up in a chase along its orbit around the Sun, *Pegasus* matched its speed with Pluto to go into orbit. Exploration of this double planet system was completed in 30 days, after which the craft set off again for the inner solar system.

On the way back the crew completed one more exploratory task: they visited a comet on its periodical return from the outer solar system. Finally *Pegasus* returned home after a total mission time of around 2240 days – a little over six years.

A Chariot to the Stars

The *Pegasus* spacecraft is the largest single structure ever assembled in space. Built at an average altitude of 483 km above Earth's surface over a five-year period, it is more than 1.3 km long (greater than 12 football pitches) and weighs almost 400 tonnes. There are three main sections to the spacecraft.

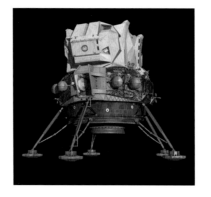

At one end a large aeroshield dominates the architecture of *Pegasus*. It's a disc 400 m in diameter, built from lightweight alloys, carbon-fibre and heat-resistant ceramics. As well as acting as a shield for 'aerobraking' in planetary atmospheres and as a thermal shield to protect the craft from the sun, it houses the main engines,

the nuclear fusion reactor, and the liquid hydrogen propellant supply. Once *Pegasus* has accelerated to cruising speed, it rotates 180 degrees so that the aeroshield faces forward, protecting the spacecraft from any interplanetary debris.

Some 700 m from the shield is the main habitation module. It has to be this far away so the astronauts are safe from the radiation produced by the nuclear engines. The module is constructed from 24 spent space-shuttle fuel tanks, ingeniously recycled from shuttle missions that serviced *Pegasus*'s construction. The habitation module provides as much internal space as 10 jumbo jets, and it houses laboratories, recycling systems for water and air, crop-growing modules, living compartments, a command centre for the mission, a viewing chamber, and storage chambers for food, water, and mechanical spares. The sleeping/exercise modules are housed at the end of centrifugal arms that rotate around the main module at up to three revolutions per minute. This produces a centrifugal force equivalent to half of Earth's gravity, making exercise and sleep easier and helping to prevent the wasting away of muscle and bone that can occur in zero gravity. An internal flywheel rotating in the opposite direction prevents the main spacecraft from spinning.

This middle section of *Pegasus* bears a series of solar panels to provide power, radiators to regulate the inside temperature, and a manoeuvrable robot arm to help astronauts working on the outside of the ship.

Above: *Hermes* – **The Io lander.**
Below: *Clyde* – **The Pluto lander.**

Introduction

The Red Team at Mission Control. Left to right:

Surgeon (Flight Surgeon) – monitors crew health. **Flight** (Flight Director) – responsible for the overall mission. **CAPCOM** (Spacecraft Communicator) – the primary communicator between flight control and the crew. The name is a throwback to early manned space flight when the first spacecraft were just capsules. **FiDO** (Flight Dynamics Officer) – responsible for aerobraking and orbital manoeuvres. **MOD** (Mission Operations Director) – the main liaison between the control room and management at the space agencies. **Guido** (Guidance Officer) – ensures the on-board navigation of *Pegasus* is correct. **PROP** (Propulsion Systems Engineer) – manages the rocket engines and orbital manoeuvring systems. **GNC** (Guidance, Navigation and Control Systems Engineer). **CIO** (Chinese Space Agency Interface Officer). **EECOM** (Emergency Environmental and Consumable Systems Engineer) – monitors the life-support systems, cabin pressure, temperature control and air and water recycling systems. **EGIL** (Electrical Generation and Integrated Lighting Systems Engineer) – responsible for all on-board power on *Pegasus*. **INCO** (Instrument and Communications System Engineer) – in charge of all the in-flight communications on board *Pegasus* and the landers. **GC** (Ground Control) – runs the global network of communications between Earth and the spacecraft. **FAO** (Flight Activities Officer) – schedules the crew's daily timetable. **RIO** (Russian Interface Officer) – is the main operations liaison between the Russian Space Agency, NASA and the European Space Agency. **MMACS** (Maintenance, Mechanical and Crew Systems Engineer) – responsible for *Pegasus*'s superstructure and mechanical systems like the truss trucks and the robot arm. **LSE** (Lander Systems Engineer) – monitors the health and operations of all the landers. **EVA** (Extra-vehicular Activities Systems Engineer) – responsible for the spacesuits used whenever the crew are outside the spacecraft. **SCIENCE** (Science Officer) – coordinates the complete programme of science undertaken during the entire mission – representing over 800 mission scientists across the world who are working on the mission. **PAO** (Public Affairs Officer) – releases mission news to the press and public. **ESO** (European Space Agency Representative).

About 300 m from the other end of the habitation module is another nuclear power plant – a Y-shaped Brayton cycle system that converts thermal energy into electrical energy using turbines. This provides additional power for the habitation module.

Planet Envoys

Pegasus carried five landers designed for visiting the surfaces of Venus, Mars, Io, Pluto and a comet. All were designed on a five-legged frame.

Orpheus made the daring descent to the surface of Venus. Protected by a tough heat shield, it plunged deep into the corrosive Venusian atmosphere and deployed a small zirconium parachute (a 'drogue') to slow its descent. It then released a large paraglider wing to carry it gently to the landing site, and it fired its rockets to make a controlled touchdown. In some ways *Orpheus* was more like a deep-sea submersible than a spacecraft. Built to withstand the extremely high pressure and temperature of Venus, its main cabin consisted of a massively reinforced titanium sphere. *Orpheus* was also equipped with a very powerful, two-stage rocket engine to lift it out of the dense atmosphere and carry it back to *Pegasus*. The external fuel tanks that powered the first part of the ascent lasted only two minutes and were jettisoned to reduce weight. The engine was then supplied by internal fuel tanks up to orbit.

Ares, the Mars lander, was also encased in a tough heat shield for its atmospheric entry phase, and like *Orpheus* it had a powerful single-stage rocket engine to get it back to orbit. But it was much bigger than *Orpheus* as it had to carry all the equipment for base camp, the rover and enough supplies for a crew of three to spend 15 days on the surface. And although Mars has much milder conditions than Venus, *Ares* still needed a strong, well-insulated cabin to resist prolonged exposure to the deep frozen low-pressure environment.

Hermes carried a single astronaut to the surface of Io, the volcanic moon of Jupiter. To protect its human cargo from the intense radiation belts around Jupiter, *Hermes* was equipped with powerful superconductors that created a protective magnetic bubble around the craft. Like an invisible shield, this bubble deflected most of the lethal charged particles in Jupiter's radiation. *Hermes* did not have to pass through a dense atmosphere, so it was less aerodynamic than *Orpheus* or *Ares*. In fact, it looked somewhat like an insect, with its systems and legs sticking out during the entire flight.

Clyde was designed to land on Pluto, the outermost planet, and was named after the planet's discoverer, Clyde Tombaugh. Like *Ares*, it was large enough to carry supplies and equipment for a 15-day stay on the icy surface, and it had the same insect-like appearance of *Hermes* because it did not have to travel through a substantial atmosphere. A nuclear reactor supplied *Clyde* with power, since the feeble sunlight on Pluto was too weak for solar panels.

Messier was smaller than *Clyde* and carried a system of tethers for its encounter with a comet. The main challenge for this lander was simply to stay on the surface after landing. Comets are far smaller than planets, and their gravity is so weak that an untethered craft could easily drift away. So *Messier* used thrusters to push it close to the surface and then fired harpoons to anchor itself in place.

Messier – the Comet lander.

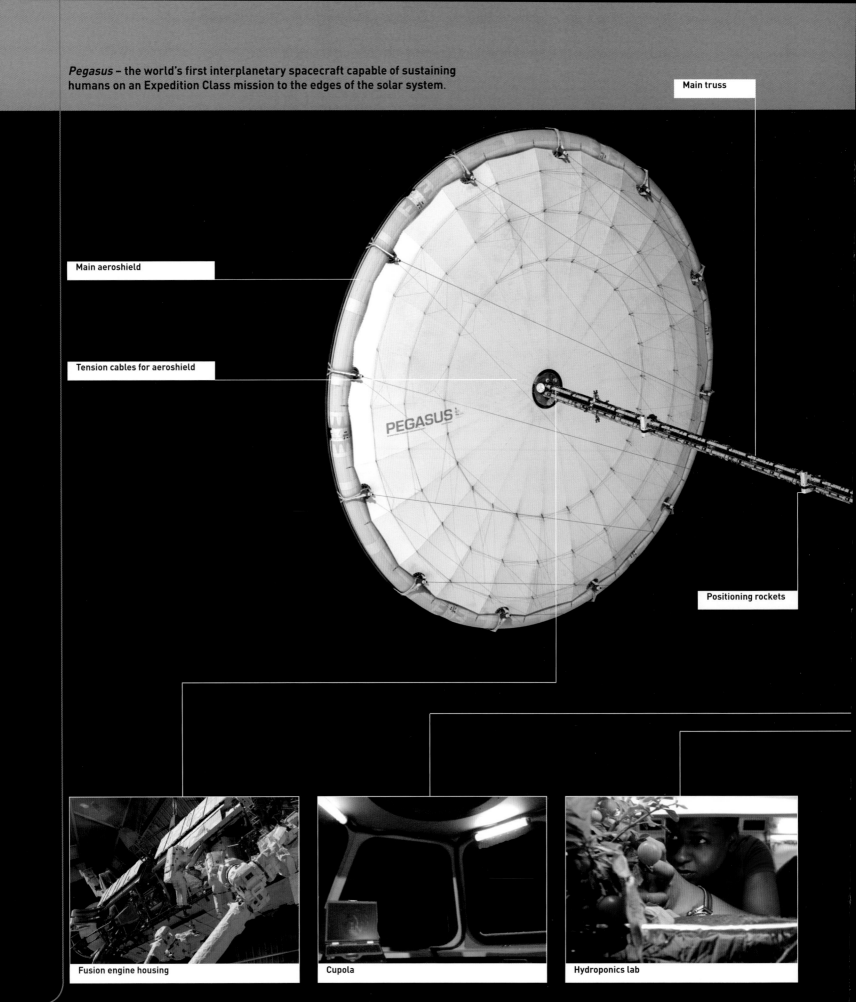

Pegasus – the world's first interplanetary spacecraft capable of sustaining humans on an Expedition Class mission to the edges of the solar system.

Main truss

Main aeroshield

Tension cables for aeroshield

PEGASUS

Positioning rockets

Fusion engine housing

Cupola

Hydroponics lab

Centrifuge sleep chamber and exercise area

Command centre

Lab zone

Robot arm

Magnetic shield generators

Brayton cyclers (power generators)

Solar panels

Radiators

Secondary truss

Lander flight deck

Pegasus airlock

Orpheus Goes to Hell

- DIAMETER: **12,104 KM (7520 MILES)** ◊ MASS (EARTH = 1): **0.82** ◊ GRAVITY (EARTH = 1): **0.91**
- SURFACE TEMPERATURE: **467 ºC (873 ºF)** ◊ SURFACE AIR PRESSURE: **90 ATMOSPHERES**
- DAYLENGTH (ROTATION PERIOD): **5832 EARTH-HOURS** ◊ YEAR: **225 EARTH-DAYS**
- DISTANCE FROM SUN: **108 MILLION KM (67 MILLION MILES)** ◊ RINGS: **0** ◊ MOONS: **0**

Tension – Alex Lloyd waits at Mission Control for news from Venus.

⭐ Tom Kirby, Commander
Mission elapsed time: 41 days, 17 hours, 28 minutes

Hanging silently above the cloud tops of an alien world is a sobering experience. Venus looks so close that you could almost reach out and touch its poisonous, yellow-white cloak of haze. Here is a world the size of Earth, but with none of the familiar features we are all so used to looking down upon from orbit. Despite the years of training, it feels strange, forbidding, even wrong to be here. The enormity of our mission has suddenly become clear.

I know this sounds odd, considering we've been flying away from Earth on a mile-long spacecraft for the last month or so. But we have all been so busy, cramming through checklists, reviewing mission rules, monitoring machines and running research protocols – concentrating so hard that none of us has really stopped to think. But after orbital manoeuvres were complete and we all took time out to crowd into the viewing chamber and

look down on the yellow-white planet below – there was Venus and that's when we all realized just what an incredible adventure we have embarked on.

We first floated through the airlock into *Pegasus* over 50 days ago now and, with the world watching, on 22 January we set sail. With one press of a button we left behind all that we knew, everyone we'd ever met, every place we'd ever been, and set off into the unknown. Hidden behind the aeroshield and our engine housing, the Earth receded. We watched on screens in the command centre. Within five hours it was a whole Earth. After six days it was no more than a big bright star with a blue tinge. We'd abandoned our direct voice communication with Earth and Control the next day. It had reached more than a minute of delay each way, and we'd switched to what would become the norm for the next six years – bursts of voice and picture communication.

Although no-one mentioned it, it was also at this time that we slipped invisibly from the gravitational clutches of Earth – the first humans to cross this boundary, the first humans to orbit the Sun separately from those left on Earth. It is a daunting thought and I don't think I could attempt this if I didn't put some of my faith in our survival with a greater force binding all this together. Falling silently towards the Sun we encountered Venus some 35 days later, growing from a small disc to the dazzling crescent that now fills our view. Each day we drew closer we noticed it seemed to take on new characteristics. The speed of this change was the only way of gauging that we were travelling more than 250 km a minute. With a single, mid-course correction burn from the engine, we hit Venus's cloud tops with pinpoint precision on 1 March, using its dense atmosphere to reduce our speed and enter orbit.

Barely a single season has slipped by back on Earth and we are already a world away – a tiny step on the scale of the solar system, but a historic one for humanity. A new world lies at our feet, and at every spare moment we stare out of the windows at the yellow-white globe. It is too bright to look at without protection, and snow blindness – or Venus-blindness, as we're calling it – is a real risk. So we all don sunglasses whenever we are in the cupola (viewing chamber) on the day side of the planet. The planet's surface is completely hidden by cloud, which makes it difficult to work out the scales of the few cloud features we can see. Are these weather patterns kilometres across or thousands of kilometres across? It is impossible to know how deeply we are looking into this haze. I thought that watching the atmospheric sounding probes drop into the atmosphere would help. But I lost sight of them long before they entered the cloud tops. Venus suddenly feels vast and dangerous. Soon two of my crew will have to go down into that lethal mist.

☒ Yvan Grigorev, Flight Engineer
46 days, 6 hours, 35 minutes

As I write this I'm in the cupola looking down on the night side of Venus. The Sun has just set (as it does every eighty minutes or so in this orbit), and below me now a gaping black hole in the stars is all that betrays Venus's presence. But if I dim the lights in here and wait till my eyes grow accustomed to the darkness, I can make out a very faint glow from the clouds. A few sheet-lightning flashes ripple faintly inside the deep clouds, but the glow is a mystery.

With so much that is still a mystery about this planet it seems all the more audacious to be attempting to land down there tomorrow. But it is of course a great honour – if we succeed, it will become a defining moment in human history, the first time man sets foot on another planet. But I cannot fool myself that anything is certain. There's so much that could go wrong. It's hard enough to land on Earth – we still get that wrong sometimes and we've done it enough times. But Venus, that's something else – 300 km/h winds, powerful electrical storms, corrosive clouds of acid, surface temperature of 485 °C, and an atmospheric pressure 90 times greater than Earth's. It's like a high-speed trip to the bottom of the ocean – an ocean of boiling acid. Not only that but we need a precision landing at the site of *Venera 14*, the last Russian robot to touch down on Venus. It all sounds impossible but I trust the technology around me completely – I have to. *Orpheus*, our lander, is the toughest piece of hardware I have seen, and I am as at home in my bulky titanium spacesuit as if it were a second skin. It's a brilliant piece of engineering. I've worn it in blast furnaces back on Earth and I've practised underwater at more than 100 atmospheres of pressure in the huge testing chambers of the Lavochkin Institute in Moscow. I trust Zoë's piloting skills implicitly. She's trained hard for this flight and I know my life couldn't be in better hands. No, I'm sure it won't be the technology that fails me tomorrow. My biggest fear is that I might not be up to the task. Maybe on that bleak and hostile planet I could just lose it somehow. I just don't want to let anybody down.

▊ Nina Sulman, Mission Scientist
46 days, 9 hours, 4 minutes

It's been hard to drag ourselves away from the windows to work since reaching Venusian orbit. After training for so long around Earth and being so familiar with every land feature, ocean basin and weather pattern, having a new planet to scrutinize from such close quarters feels like an incredible treat.

As I stare out at this dazzling, yellowish-white orb, I can't help but wonder what unknown life this planet perhaps once harboured. There's every reason to believe that Venus might have been like Earth early on in the evolution of the solar system. Perhaps primitive life got going here just as it did on Earth. And although the surface is too extreme for life today, it's possible alien microbes might still cling on, floating in the haze above. In fact, the presence of microbes in the atmosphere would neatly explain some intriguing observations of Venus's atmosphere from previous spacecraft.

The Mysterious Evening Star

Bright enough to be seen in the daylight up to three hours before sunset or after sunrise, Venus was obvious to the first civilizations, who claimed it as their Goddess of love as early as 3100 BC. As our nearest planet, it was, until recently, the third brightest object to be seen from Earth after the Sun and Moon; the *International Space Station* has now knocked it into fourth place. Despite Venus's brightness, the first telescopes could see little more than the top of its dazzling, thick atmosphere. In fact, until the 1960s, we weren't even sure which way it spun, how long its day was, or if it had a hard surface. Here, then, was a planet we had known of for centuries but which we knew almost nothing about.

Galileo Galilei, Professor of Mathematics at Padua, Italy, was the first to magnify Venus – with his 'optick tube' –

and was quick to note its phases, proving, controversially, that it orbited the Sun. Later astronomers concluded from the apparent uniformity of the planet that it must have an atmosphere, and in 1761 the Russian astronomer Lomonosov published accounts of an arc of light outlining the night side of the planet. He reasoned that this was light from the dayside being refracted through an atmosphere at least as dense as Earth's. He could never have guessed at just how dense it was.

During the next two centuries, improvements in optics prompted both a search for physical details in Venus's face and attempts to work out the chemical composition of the atmosphere. But other than charting a few vague cloud features and discovering that the atmosphere was largely carbon dioxide, astronomers added little to our knowledge of Venus. In fact, the failure of optical astronomy to reveal more about our nearest planet led to some extravagant suggestions about what might lie beneath the clouds. Ideas ranged from a moist, steamy world of luxuriant vegetation and primitive animal life to vast oceans of oil. Some claimed the planet was covered with water, while others suspected it was a gloomy desert with violent gales that levelled everything in their path.

Venus needed a new branch of astronomy to unveil its secrets. In the 1950s, astronomers discovered how to pierce the thick clouds by using a technology pioneered in the Second World War: radar. By bouncing radar signals off Venus and analysing radio emissions from the planet, astronomers worked out that Venus had a rocky, dry surface heated to

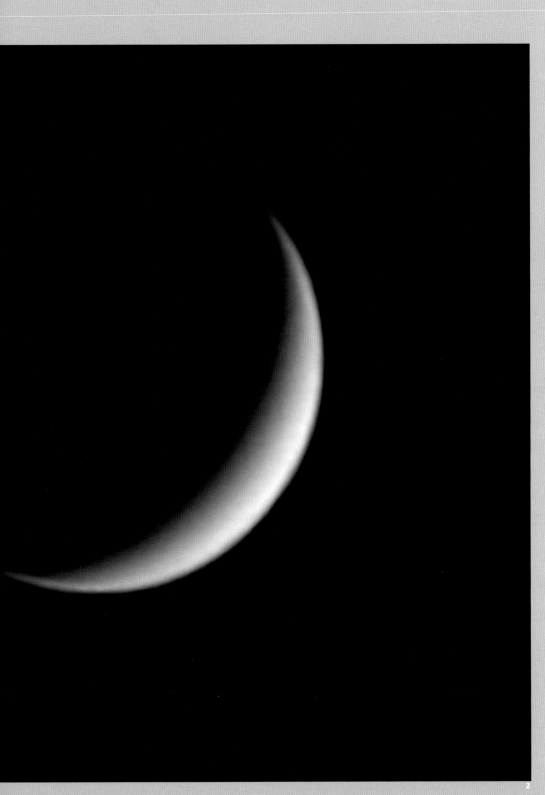

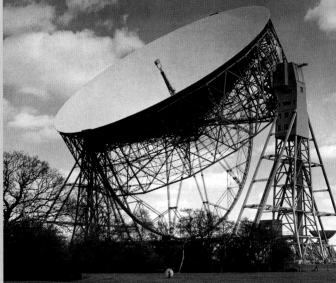

more than 400 °C (750 °F). The planet span very slowly, taking 243 Earth-days to complete one Venusian day – which was longer than a Venusian year (225 Earth-days). Stranger still, Venus spun from east to west as if it were upside down. But these tantalizing facts were controversial, and there was really only one way of confirming them – we had to go there.

1. Galileo Galilei – one of the first astronomers to magnify Venus with a telescope.
2. A crescent view of Venus – magnified with a modern telescope.
3. Radio telescopes like the Lovell telescope in Cheshire can penetrate Venus's thick clouds to reveal the surface.

What we need is a sample of the upper atmosphere to analyse back on *Pegasus*, and so, on our return to orbit tomorrow, that's exactly what *Orpheus* is going to attempt to snatch through a gas-sampling device in the nose cone. The scientists who first made the suggestions of microbes accept that nobody will believe there's life on Venus until we see it under the microscope and it's moving and waving back. By this time tomorrow, I'll be looking to see if anything is waving!

Lessard – preparing for descent.

💬 Zoë Lessard, Mission Scientist
47 days, 5 hours, 40 minutes

The long wait is about to begin. I've always known this was going to be the toughest part for me. I can hear Yvan in his chamber getting into his suit, but all I can do is stare at the instruments and pray nothing goes wrong. So far *Orpheus* has behaved like a dream.

The descent was a rough ride but there were no hitches. We undocked from *Pegasus* just over 90 minutes ago, and the first

Chapter 1

Grigorev gets a rough ride.

Orpheus drops away from *Pegasus* – commencing its journey into hell.

visible signs of descent were when the black skies of space were tainted by wisps of yellow haze from Venus. At 100 km altitude we could see the acrid yellow atmosphere stretching to the horizon through the capsule's window, and by 70 km up the clouds had grown to be a dense haze. Outside the temperature was a comfortable 13 ºC and the pressure was about the same as on Earth's surface, but I knew this would change as we plummeted down – *Orpheus*'s thick walls were engineered to withstand more than 150 atmospheres. I remember waiting for the Gs to start ... right on cue, I felt a growing compression around my back – like a big hand pulling me up as *Orpheus* was slowed down by the drag of the thickening atmosphere. Several unstowed items I'd not

noticed dropped abruptly to the floor of the capsule. The Gs doubled and then tripled. The greatest force on my body that I'd experienced in the past six weeks was half a G in the sleep compartments, so what Venus was throwing at us felt crushing. I could only describe it as like having several people sitting on my chest. My breathing was reduced to short puffs and grunts as I tried to keep in voice contact with Tom.

The air speed indicator was touching 19,000 km/h when I glanced over at the heat shield temperature indicators – they were already reading more than 2000 ºC. Orange flames flared outside the tiny window, obscuring the view of the clouds. Transmission to

Grigorev – shortly before getting into his pressure suit, which can be seen behind.

Pegasus was starting to get patchy, and as I lost contact I continued to report progress to Yvan over the intercom. I could just make him out through a small window to my right, sitting in his sealed airlock. He couldn't see much and had kept quiet so far.

The next six minutes of radio silence felt like six hours, as the outside of *Orpheus* turned into a fireball, ionizing the surrounding air and halting radio communication. As we came out of the high-temperature phase of entry, I switched to a different radio antenna to reacquire *Pegasus*. I didn't pick up Tom's voice until over seven minutes had elapsed. Although faint, I could hear the strain in his unemotional delivery. I reported all systems normal and remember trying to look relaxed in case picture transmissions were getting through too.

Chapter 1

At 55 km altitude, the heat cover was blown from the top of the capsule and a small extraction parachute caught the yellow air and tugged a tough zirconium parachute from its cocoon on top of the capsule. The zirconium 'chute slowed the craft further, bringing

our speed down to less than 1600 km/h. During this rapid deceleration the G-force momentarily surged to 8 and I was unable to speak or breathe. My vision disappeared at the edges. I blinked to reassure myself that my eyes were OK and strained to read the instruments and watch the automatic landing procedure unfold. Next up was the main parachute deployment. *Orpheus* needed to slow down to a comfortable 10 km/h to survive touchdown, but we were still plummeting at more than 1000 km/h. There was a sudden kick to the outside of the capsule as four little rocket engines blasted the heat shield off the bottom of the craft. On a monitor I watched it tumble away into a layer of cloud below. Overhead, the zirconium parachute was jettisoned and a giant paraglider wing was deployed. It opened rapidly in the dense air, slowing the capsule even more, before a powerful wind tugged us sideways through the yellow sky. The computer turned us and started steering towards a point just east of Phoebe Regio – a vast lava plain about 50 degrees west of the Venusian meridian. By now our vertical speed had dropped and condensation began to stream off the window. This exotic Venusian moisture wasn't water but a cocktail of hydrochloric, hydrofluoric and sulphuric acids collected from the thickening clouds that the capsule was passing through. But the moisture was short lived. Within another kilometre the temperature had risen too high for liquids and the window boiled dry.

Below us loomed an opaque layer of cloud: Venus's densest cloud zone, which lay about 50 km above the ground. We crashed into it – and all my views were obscured. Visibility was briefly down to less than that in an Earth fog. The Sun disappeared, showing up as nothing more than a vague glow overhead. Then we passed out of the dense cloud and into a clearer zone – we were approaching the scalding lower atmosphere. A sudden bright flash outside distracted me. It wasn't lightning. It was much brighter. Yvan had seen it too. It signified a significant milestone in our descent a single metre-thick band where the atmosphere goes supercritical, simultaneously taking on the properties of a gas and a liquid. We were 20 km from the ground, cockpit readouts showed the atmospheric pressure was already 40 times that on Earth and the temperature had climbed to more than 350 ºC. A sickly yellow light now suffused the inside of the cabin.

At this point, *Pegasus* informed me that the sounding balloons we had deployed before our descent were detecting higher than average amounts of sulphur in the atmosphere and stronger than expected vertical winds. This hinted at the possibility of a recent volcanic eruption in the vicinity of the landing site. The geologist in me was excited by this news, but I knew as an astronaut about to land on the surface, it was bad news. I heard Yvan groan.

The monitor showed little below us other than a deep yellow fog, with alternate dense and lighter layers a few tens of metres thick. But at exactly 7.2 km above the ground, the haze cleared and a spectacular vista flashed onto the screen. The colour suddenly shifted to a strong orange-yellow, and at the same time the temperature jumped another 100 degrees, taking it to more than 450 ºC. Noisy air-conditioning systems kicked in immediately, straining to bring down the rising temperature inside *Orpheus*. Outside the capsule I could see the extractors glowing as they vented off the excess heat. We would not survive a minute if they failed.

Tom Kirby's reassuring voice confirmed that we were into the superheated surface zone. It was our first view of the ground and the landing site. It was just like the radar images, and thankfully there was nothing that looked like a volcanic eruption in progress. There was another sharp thwack to the capsule as the giant paraglider canopy was jettisoned and retrorockets kicked in beneath the lander to control our descent. The ground looked

rubble-strewn as we got close, but hardly any dust was disturbed by the rockets. I could see things remarkably clearly, given the density of the atmosphere we were immersed in.

Finally, *Orpheus* bumped and shuddered and a contact light flashed up – we'd made it. After a few hurried post-landing checks, I reported our success to *Pegasus*. 'Panina Base here … *Orpheus* has arrived in Hades.'

It's over to Yvan now. He has 80 minutes to conduct his 'EVA' (extravehicular activity), before the suffocating pressure and temperature will begin to overwhelm *Orpheus* and our chances of a successful launch evaporate.

👁 Zoë Lessard, Mission Scientist
47 days, 6 hours, 35 minutes

Almost an hour has elapsed since Yvan stepped outside. We're running late and I can hear he is exhausted out there.

Until Yvan left, I'd been too busy to stop and think about where we really are. But now I'm alone with my thoughts and able to contemplate the extreme nature of this place. With the engine off, all I can hear is the wind blowing around the outside of the capsule, and every creak and groan brings fear of structural failure. This is no simulator. This is for real. My paranoia drives me to start examining the inside of the cabin for cracks. There's no room for a single flaw when almost 900 tonnes of air is trying to force its way into every square metre of exposed spacecraft.

I've been staring through the tiny window, transfixed by the extremeness of this place. But it feels like the poisonous world on the other side of the glass is draining my life. We're not supposed to be here. And with every moment we stay, the chance of a successful launch decreases. Throughout the EVA, *Orpheus*'s systems have been suffering in the crushing heat and pressure, some failing altogether and others limping on. Most are designed to bypass failures – in a sort of self-repair mechanism. But there is one system that can't fail safely: our ascent engine. If that dies then so do we.

👁 Zoë Lessard, Mission Scientist
47 days, 8 hours, 50 minutes

It was such a relief when I saw Yvan's helmet bobbing into the hatch again and I watched him slump into his seat, exhausted.

Grigorev begins his exploration of the landing site.

The tensest hour of my life was over, and we wasted no time preparing *Orpheus* for takeoff.

At 21.42 I lifted the switch guard and flipped the ignition switch. Nothing! Banishing thoughts of failure I reset the system and flipped the ignition again. The explosive bolts on the leg base released us, and with a huge kick we began our ascent. Outside I could see huge plumes of brown steam billowing up into the orange sky. Bless her titanium hide, *Orpheus* was blasting us away from that grim landscape. As we climbed up into the murk again, the roar of the dense atmosphere rushing around the outside of the capsule became so loud that it almost drowned out the engines, but I could still hear Yvan shouting in encouragement through the din.

We needed to reach almost 25,000 km/h to escape the gravity of Venus – nearly as fast as a rocket launch from Earth, and through a much denser atmosphere. The ascent felt smoother than the descent because the engine matched our speed to the density of the atmosphere, preventing us from being torn apart. At two minutes into the climb there was another kick as we jettisoned the external fuel tanks and they peeled away into the gloom. We were free from this suffocating planet and heading home.

Light levels grew throughout our ascent, but I was still surprised when we cleared the last of the clouds and the skies were suddenly black again. 'Catch up' with *Pegasus* took a further hour of delicate manoeuvres. I was surprised at the joy I felt when I saw her for the first time – the large, familiar disc and her skinny, insect-like body meant home and safety.

Grigorev looks out onto a grim planet.

Alien Life in the Clouds

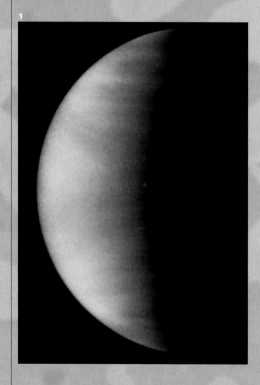

Although the idea of life on Venus has never been as popular as the idea of life on Mars, there was one 'scientific' theory that held sway for much of the twentieth century. In 1918 the Swedish astronomer Svante August Arrhenius detected clouds in the atmosphere of Venus and decided that clouds meant water. It seemed reasonable to suppose that the wet atmosphere might conceal a wet planet – perhaps even a world of swamps and jungles. For over four decades this fantasy fuelled the stories of science-fiction writers and the curiosity of astronomers. But in the 1960s came proof that the surface was a hot, dry place, and thoughts of life evaporated with the water.

Dry as the surface of Venus undoubtedly is, there is still a trace of water in the atmosphere – not as much as Arrhenius had supposed, but it might be enough to prop up an alien ecosystem. This theory was given renewed support in 2003, when scientists re-examining space-probe data discovered several anomalies in the chemistry of the Venusian atmosphere. The anomalies, they claimed, could be explained by microbes harnessing the Sun's energy and feeding off sulphur in Venus's upper clouds.

Chemistry predicts that solar radiation and lightning in the top of the Venusian atmosphere should produce large quantities of carbon monoxide – yet there's hardly any there. It's as if something is removing it, says principal investigator, Dirk Schulze-Makuch. What's more, probes have detected both hydrogen sulphide and sulphur dioxide, two gases never normally found together because they react with each other. Something must be producing them. Stranger still is the presence of carbonyl sulphide, a gas so difficult to produce inorganically that it is sometimes considered a definitive indicator of biological activity. Schulze-Makuch concedes that there may be nonbiological ways of producing hydrogen sulphide or carbonyl sulphide, but the most efficient way of making them on Earth is with microbes.

He suggests that bugs in the Venusian clouds could be combining sulphur dioxide with carbon monoxide and possibly hydrogen to produce either hydrogen

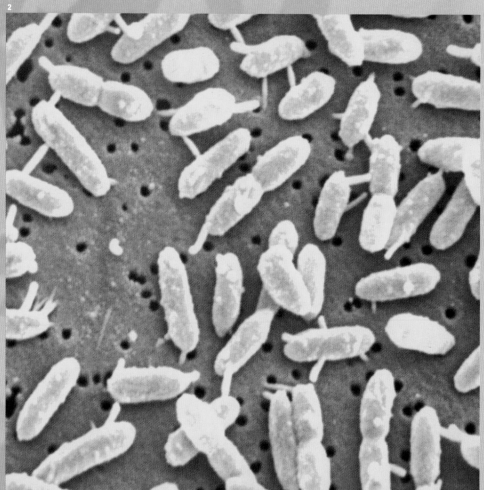

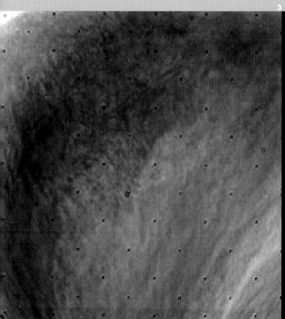

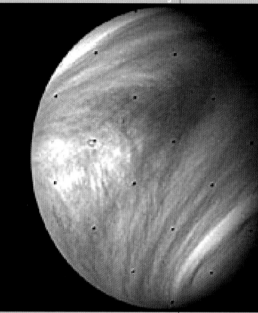

sulphide or carbonyl sulphide in a metabolism similar to that of some early Earth bugs. The bugs could be using ultraviolet light from the Sun as an energy source. And if they are absorbing UV, that might explain the mysterious dark patches on UV pictures of Venus taken by *Mariner 10*.

Conditions in the atmosphere at around 50 km (30 miles) up are relatively hospitable – the temperature is about 70 °C (160 °F) and the pressure is about 1 atmosphere. And although the clouds are very acidic, this region has the highest concentration of water droplets in the Venusian atmosphere. 'From an astrobiology point of view, Venus is not hopeless,' says Schulze-Makuch. 'Look at Earth's own atmosphere. There are microbes growing and reproducing there and, by adopting one or more survival strategies, why not Venus.' One adaptation

the microbes might employ is to coat themselves in a thin layer of sulphur compounds to screen out the worst of the Sun's radiation. Such a cocoon would also protect them from acid in the clouds.

Critics are not convinced and point out that life needs much more water than the few tiny droplets in the Venusian skies. Schulze-Makuch plans to turn the Hubble telescope to Venus to make further observations. Ultimately, though, a sample from the upper layer is needed to determine whether these chemical anomalies really are a signature of life.

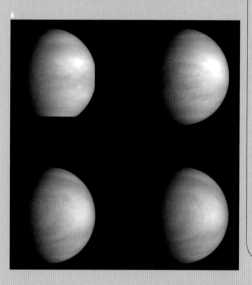

1. An ultraviolet view of Venus from the Hubble telescope.
2. Acid-loving extremophiles bacteria – a possible candidate for chemical phenomena detected in the Venusian atmosphere.
3–5. The *Mariner 10* spacecraft's view of Venus - snapped en route to Mercury.

6. Four violet-light views of Venus taken by the *Galileo* spacecraft to reveal cloud circulation patterns.

Orpheus lifts off. The camera that took this picture disintegrated moments later.

⊞ Tom Kirby, Commander
47 days, 9 hours, 40 minutes

Space-flight experience has taught me never to be complacent. However well designed and built our spacecraft are, there are no guarantees that everything will go to plan. So right up until *Orpheus* had hard docked I found it hard to relax. And when Zoë and Yvan floated through the hatch, relief was my overriding emotion. But what a triumph; so many people had told us a human visit to this planet was utterly impossible, and yet we'd accomplished it. Even so, Yvan looked exhausted. The last three hours had certainly taken their toll on him. He was ashen and still perspiring. But he was clearly buoyed up by his success and was even cracking jokes about the smell. He'd noticed the stench of sulphur within a few moments of the EVA and he guessed he'd brought it back with him. He was right. He stank. In fact, everything that had come back reeked of sulphur, and for a few hours the stench of Venus was overwhelming through most of *Pegasus*. The first human to walk on Venus needed a good wash!

During the three hours they were away from *Pegasus*, I was more conscious than ever of the vast distance between us and Control. Getting a reply to radio messages now took six minutes, so if *Orpheus* had needed emergency help, it would have been up to me to make decisions – I had been solely responsible for bringing them back alive. Everyone at Control knew this and kept quiet during most of the mission. I was only aware of them when the room erupted in celebration six minutes after Zoë announced their safe landing! It was the first descent I'd commanded from orbit without the help of Earth. But I'd only realized how heavily this responsibility had been weighing on me once it was lifted and they were safely back on *Pegasus*.

⊠ Yvan Grigorev, Flight Engineer
49 days, 6 hours, 35 minutes

For most of the extraordinary hour I spent walking on the surface of Venus, I was too focused on the timeline of the mission to stop and think about where I was. It is only now – a couple of days after returning from that hellish place – that it is beginning to dawn on me what we did.

It was a gruelling ride down, but Zoë piloted us right on target. I sealed myself into the titanium suite and made a last check of the airlock conditions and my life-support systems and coolant supplies. The temperature and pressure inside my capsule had to match the outside before I could open the hatch. Sealed in the suit, I was oblivious to just how lethal the environment was becoming in the airlock around me.

I remember looking at the image from the camera on the lander legs. The camera was to relay this historic moment to the world but, to my disappointment, it had failed as I had opened the hatch. Great start, I thought. My first footsteps would only be caught by the cameras on my helmet. There would be no other pictures until I got the first of three cameras erected on the surface.

At last, the airlock was ready and the hatch opened. Inching backwards out of it, I felt for the foot holes below the door. This was by far the most treacherous part of my descent – I didn't want to fall from this height! Eight steps took me down to the lander leg and from there it was another five rungs to the footpad. Aware that no-one could see my progress, I relayed a continuous commentary to Zoë and Tom.

On the footpad I swung my oversized right foot forward onto the flat pavement of brittle looking lava and stepped onto Venus. It was 20.45 universal time. 'I'm on the surface,' I muttered almost to myself. Without time to take in the enormity of what I'd just done, I headed out onto the lava plain, taking a second and third and fourth step. All I was thinking about was my first task. I continued the running commentary for the geologists. The dense air offered resistance as I moved through it and the ground was hard under foot, but it gave way slightly in places when I pressed hard. It was a brittle lava of some kind – slabs of rock that looked like basalt – and there was almost no soil. The slabs consisted of many closely packed horizontal layers, each about 1–2 cm thick. They seemed very dark, but in the orange light their exact brightness was hard to determine. The light really did match a dull winter's day in Moscow, just as the Soviet Venera probes had discovered decades ago. There were absolutely no shadows, and the yellow-orange light seemed scattered uniformly, creating an eerie twilight.

I remember thinking how calm it felt after that bumpy descent and landing. There seemed to be no wind, which was strange considering how turbulent the upper atmosphere was. The Sun was hidden by cloud, but an ill-defined glare to my west showed where it was. It was morning – and of course it still would be for another six weeks.

Nonidentical Twins

Imagine a world where every day is dull and overcast. The Sun, could you see it, crosses the sky 100 times more slowly than on Earth. In fact, it takes two weeks for dusk to fall, and the Sun sets in the east (the opposite to Earth). Sunset is followed by an interminable night that lasts longer than one of our seasons. It's so hot that lead and zinc would form rock pools, and mountain peaks get dusted with a metallic 'snow' of iron pyrites and germanium. Welcome to Venus.

This is a planet of comparable size to Earth, and which formed in a similar part of the solar system, from the same chemicals, at the same time. Yet a human on the surface of Venus would fare about as well as the proverbial snowflake in hell.

Our fundamental differences stem from the hand dealt to each planet by the chaos of the early solar system – a place populated by tens of rocky worlds in crisscrossing orbits. Collisions were probably common, and within the first billion years of their formation, both Earth and Venus underwent catastrophic impacts that destroyed their original crusts and shifted their axes. Venus's axis was probably pushed right over in a smash that almost halted the planet's rotation, accounting for its slow westward spin today. Earth's impact also left its mark, tilting our axis to 23.5 degrees but leaving us with a rapid eastward spin and a moon.

The Sun's power was weaker in the early years of the solar system, and Venus would have been cool enough for water to exist as a liquid on the surface. But as the Sun grew hotter over the next billion years, the zone in which liquid water can exist was pushed further out into solar system. Venus was baked dry, and much of its water was lost into space. Earth, being further from the Sun, stayed in the wet zone. With no oceans to soak up carbon dioxide gas from volcanoes, Venus fell victim to a runaway greenhouse effect. Carbon dioxide levels rose, trapping the Sun's energy and driving up the temperature still further. Earth, in contrast, had oceans and a biosphere, which could absorb carbon dioxide as quickly as it was produced, helping to maintain the planet's moderate temperature.

Earth's water had another important effect. It kept the crust flexible enough to sink back inside the planet and remelt. As a result, Earth's crust is made up of mobile plates that slowly drift across the planet, and heat from the interior can escape between the boundaries. In contrast, Venus seems to be enclosed by a single, planet-sized plate of rock and heat, trapped inside escapes by conduction or through the abundant volcanoes that punch their way through the crust.

There is much we still don't understand about how Earth and Venus became so different, but their distance from the Sun seems to have been the ultimate cause. As our star's heat grows over the next two billion years, Earth may yet lose its water, and then rock pools of lead may grace our dry beaches too.

Top left: A portion of Venus's surface unveiled through its thick clouds by the Arecibo radio telescope in northwest Puerto Rico.

1–3. Three faces of Venus – unveiled by NASA's radar mapping mission Magellan – which charted almost the entire planet during a single Venusian day in the early 1990s. The false colours relate to the topography of the surface.

Where Mountains Sink

Vast plains cover more than 70 per cent of Venus. Between the plains are two main areas of highland: Ishtar Terra in the north, and Aphrodite Terra straddling the equator. These are sometimes referred to as continents, but are not separated by oceans.

Ishtar Terra (named after the Babylonian goddess of love) is the size of Australia – about 2500 km (1550 miles) across – and sits in a position equivalent to Earth's northern Atlantic. At its western end is a high, lava-coated plateau called Lakshmi Planum, flanked by steep slopes. It rises to an altitude of 3000 m (9840 feet) – comparable in size to the Tibetan Plateau. (On Venus altitudes are measured from the mean radius of the planet. The equivalent point on Earth is 3 km (2 miles) below sea level, so to compare Earth's altitudes to those on Venus you must add 3 km (2 miles) to everything.) Ishtar is dominated by the great Maxwell Montes – a chain of giant peaks rising to an altitude of 11,000 m (36,000 feet). East of the Maxwell mountains is a lower plateau as large as western Europe. It consists of rather broken terrain that has been termed *tessera*, the Latin for tile, because it resembles a vast mosaic floor.

Aphrodite Terra (named after the Greek goddess of love) is larger than Ishtar, at more than 10,000 km (6200 miles) long and 3000 km (1900 miles) wide. If superimposed on Earth's equivalent latitudes, it would stretch from the Indian Ocean to the southern Pacific. It is shaped like a giant scorpion, its west-facing head made up of the higher terrain of Ovda and Thetis Regio, and its tail formed by a series of canyons that arch up to the east and north. The

longest of the canyons is Diana Chasma, which is some 1000 km (620 miles) long and over 4000 m (13,000 feet) deep in places.

Venus also has two small highland regions. The smaller of the two is Alpha Regio, and to its southeast is Beta Regio, with its large and probably active volcano

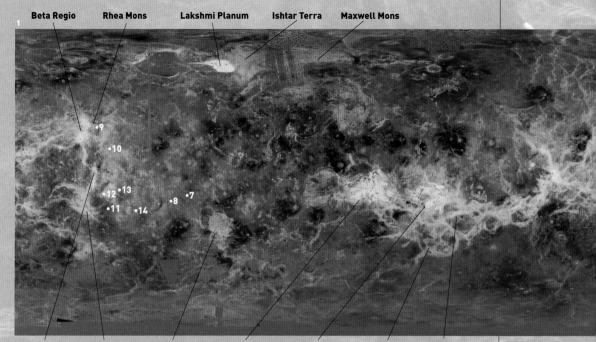

Top labels: Beta Regio · Rhea Mons · Lakshmi Planum · Ishtar Terra · Maxwell Mons

Bottom labels: Theia Mons · Phoebe Regio · Alpha Regio · Odva Regio · Thetis Regio · Aphrodite Terra · Diana Chasma

Theia Mons, which stands in the shadow of an even bigger mountain, Rhea Mons. The constant lightning over this region suggests it may be volcanically active.

Lowland plains cover most of the rest of Venus. The Venera landers discovered that these plains are littered with plates of basalt – a volcanic rock that forms when lava cools quickly. These plains all seem to be of a similar age – around 500 million years – suggesting that they all formed together around the whole planet,

producing a very different style of geology to the Earth. Dotted among the plains are craters, fissures, complex wrinkled areas called arachnoids, and vast, mysterious volcanic structures 100–1000 km (60–600 miles) wide, called coronae. Coronae might be caused by plumes of semimolten rock rising from deep inside the planet. On Earth this process would form a mountain, but on Venus the mountain seems to collapse back into the hot crust. It's a mystery why some mountains sink while others, like Maxwell Mons, stay standing.

1. Mercator projection of the Venusian surface. Our nearest planet is about the same size as Earth, but with no oceans it has more than three times the land. The Venera landing sites are marked in white. (See pages 42–3.)

I glanced at my environmental monitors to check the early-morning temperature and pressure – it was 486 °C, and the air pressure was 90 times greater than on Earth, or the equivalent of being half a kilometre beneath the surface of the sea. Somehow I found this reassuring – it was exactly what was predicted. The temperature and pressure on Venus don't change much during the day or night. Always gas mark 27.

Microphones on the outside of the suit were feeding me ambient sounds, and for the first time I became conscious of intermittent, loud claps of noise. It was hard to tell, but I think they were coming from the distant Beta Regio mountains. There was certainly a lot of electrical storm activity there, and I assumed it was thunder. Although I couldn't see lightning I could hear it causing interruptions in the communications link with *Pegasus*. We knew that the dense atmosphere holds a lot of charge, stopping some signals altogether and distorting, bending or deflecting others. But it was worse than we'd anticipated. I stopped listening to Tom back on *Pegasus* and just concentrated on Zoë's voice.

A checklist display inside the helmet described my detailed programme of activities minute by minute. The next task called for a contingency sample to be taken in case we had to abort the EVA prematurely. I strode towards the surface-science package that had been deposited beneath the third leg on landing. Inside was an aluminium rock box. I spent the next six and a half minutes working hard to fill it with broken bedrocks and other loose material. Those samples should help to date the surface and reveal when the volcanoes of Panina were last active.

The second task was to set up the camera units, giving *Pegasus* and Control their first good look at the locality. I asked Zoë if she could see anything. All three cameras were functioning well at first, but one of them packed up after just a few minutes. I lumbered over to see if I could fix it quickly. There was not much left to fix. It had imploded, part of the insides quickly melting in the heat – a little reminder from Venus.

Grigorev deploys a seismic sensor network on the surface of Venus at the landing site.

Grigorev approaches the Russian *Venera 14* spacecraft – the last robotic explorer to reach the surface of Venus in 1982.

Next task was to deploy the seismic station. One of our mission goals was to check for volcanic activity on the planet, and given the anomalous sulphur readings in the area, there were high expectations for this experiment detecting something. The geologists had their money on Theia Mons, a mountain 3200 km to the northwest that showed signs of recent volcanic activity. Positive seismic readings from Theia (or from anywhere else on the planet) would go a long way to solving the arguments of planetary geologists back on Earth.

The seismic network took nine minutes to lay out and install properly. It proved harder getting the contact teeth embedded in the ground than it did in training. I understand that it is still working, though it has been disappointingly quiet since it registered *Orpheus*'s launch. We hope it will last at least a month – maybe longer.

Venera 14.

My schedule was slipping badly, and Zoë advised me to head towards *Venera 14*. According to our best estimate, the Soviet probe should have been within walking distance, down slope and to the east of *Orpheus*. She landed here back in 1982 and sent back some of the first colour pictures of the surface. There was little on *Venera* that could melt or crack, so the engineers reckoned she'd still be intact, except for a bit of chemical corrosion and bleaching and a good coating of dust. My task was to bring back part of *Venera*'s titanium–aluminium body so we could see how prolonged exposure to the Venusian atmosphere had affected it.

Sure enough, I could see the dark silhouette of something where *Venera* was predicted to be. I estimated she was about 50 metres away, but it proved hard to judge distances in the heat-rippled Venusian air. What little sunlight reached the surface was refracted by tiny changes in temperature in the super-thick air, distorting the view. I headed for where I thought *Venera* was, but a minute later I was still not there. It was further than I thought. The dead calm of my first few minutes on the surface had now passed, and gusts of wind were beginning to throw me off balance. The wind speed was slow, but the air was so dense that it packed a punch five to six times stronger than on Earth.

After an exhausting walk in the heavy suit, I saw *Venera* emerge clearly out of the gloom. I know we were expecting it to be there, but to be suddenly standing next to it seemed unreal, hard to believe. It looked just as it did in the pictures taken here all those years ago – there was the analyser arm resting on the lens cap. I kicked the cap out from under the arm with my left foot and rested the detector arm on the rocks it was designed to study. I searched for the base plate and its blocky insignia 'CCCP', which declared the probe's Soviet origins. But the base was covered in a thick coating of Venusian dust.

It took a lot of effort to take a sample from *Venera*, and even before I started back I was exhausted and overheating. When I turned to look for *Orpheus* I had a moment's panic – the huge lander seemed so far away, and it was uphill. I really wasn't sure I had the strength to get there. Zoë recommended dropping the extra weather experiment and just heading for *Orpheus* with the samples. Every moment of that walk was agony. Every muscle was burning, and Venus seemed to have conjured up a headwind. At one point I stumbled while trying to pick up the sample box. Contact with the hot ground meant that, for the first time, I felt the temperature rise inside the suit, and I heard the life-support system kick in to compensate. I took strength from this; the suit was working fine and there was no reason for me to fail. I got back to my feet with a little help from a nearby camera stand and, after stowing the sample box, dragged myself up the ladder and into the airlock.

Orpheus's legs remained on the surface at take off, and fixed to them is a plaque recording our visit. Its inscription reads 'Here an explorer from Earth first set foot on Venus, carrying the human spirit of exploration to the next frontier.' Below this are our names and signatures, and above the name of our trusty little lander.

It still doesn't feel like we did it. Even when I play the video recordings, I feel like I'm watching the mission through someone else's eyes. But we did, we did do it. And the world was watching.

A tiny imperfection in the casing spells doom for one of the surface cameras.

➕ John Pearson, Flight Medic
50 days, 9 hours, 13 minutes

Yvan has recovered well from his EVA and seems to have suffered no long-lasting effects after his ordeal on the surface of Venus. He and Zoë slept for well over 12 hours during the rest period following their mission, and they've been consuming more than the allocated 2700 calories a day food ration. I was particularly concerned that Yvan had become badly dehydrated as he struggled back up the slope to *Orpheus*. I know it disappointed the scientists back on Earth, but dropping those extra experiments might have saved his life. His spacesuit weighs 91 kg, and although he'd been training throughout the voyage to Venus, it was a Herculean task to reach *Venera* and get back to *Orpheus*. Understandably, he's still in high spirits following the huge success of his efforts and the place he's gained in history. In fact, we have all been boosted by their success. Control have been

sending us web pages and other media headlines heralding Yvan as the man of the moment and singing the success of *Pegasus*. Sales of Yvan dolls have rocketed. I've allowed him time to do a series of press interviews and educational outreach events with schools. I know he wants to do more, but I'm mindful of overloading him. There's still a full daily maintenance programme for him to keep up with on *Pegasus*, and we've got a long way to go.

We're less than a fortieth of the way through our marathon expedition, and I've warned him of the low that might follow his present exuberance, when the press and public interest dies away and he realizes how much further we still have to go – much of it exceedingly mundane compared to the exhilaration of the past few days. I don't want to seem like a killjoy, but it's for his own good.

Descent into Oblivion

Close to the Sun, with a hot, super-dense atmosphere, Venus posed immense technical challenges for visiting spacecraft. But tenacity and determination, fuelled by cold-war rivalry, ensured that our nearest planet was also the first to be visited by our robots.

A mission to Venus was the next obvious demonstration of superpower prowess once exploration of the Moon was underway in the early 1960s. Keen to maintain their superiority, the Soviet Union launched the first Venus-bound spacecraft – *Venera 1* – on 4 February 1961. But this hurried attempt, and at least four more that followed, all failed to reach their destination.

With the prize still up for grabs, the USA fast-tracked the development of a series of lightweight Venus probes called the Mariners and, by 1962, two of them were ready for launch. On 22 July *Mariner 1* lifted off from Cape Canaveral – but it didn't get much further, plunging into the sea for want of a minus sign missed off its guidance software. Hurried modification of *Mariner 2*'s launch rocket saw a successful liftoff for the twin and, against all odds, after a nail-biting, four-month journey, it flew past Venus on 14 December. As it disappeared around the far side of the planet, its radio signals passed through the planet's atmosphere – confirming surface temperatures of more than 400 °C (750 °F) and an atmosphere of almost pure carbon dioxide. There were definitely no primeval swamps down there.

Four days after its Venus encounter *Mariner 2* broke down as it succumbed to robotic sunstroke.

Russian intelligence suggested the USA was to launch another Venus-bound Mariner probe in June 1967; beaten to the first fly-past, the Russians decided to attempt a daring landing on the surface. Two days before *Mariner 5* blasted off, *Venera 4* was launched. When it reached Venus it released a landing capsule that parachuted through the atmosphere for 94 minutes before contact was lost. The Russians convinced themselves it had landed and claimed another space-race first for Communism. But the truth was that it failed catastrophically when the temperature topped 280 °C (536 °F) and the pressure climbed to 22 atmospheres. It had been crushed while still 25 km (16 miles) above the surface.

A successful surface landing would take a lot more ingenuity. Two more Venera failures, crushed on descent, led to further redesigns. Ten days before Christmas 1970, mission scientists held their breath once more as *Venera 7* entered the atmosphere. Frustratingly, contact was again lost during the descent, but after scientists re-examined the data they found that conditions around the probe had stabilized at 475 °C (887 °F) and 90 atmospheres – suggesting that the spacecraft had come to rest. It was the first signal to be transmitted from the surface of another planet.

By 1975 the big prize for robotic space exploration was still untaken – nothing had successfully landed on another planet and returned a picture. *Venera 8*, though successful in reaching the surface of Venus and confirming the severe conditions, hadn't carried a camera. Both superpowers had lacked the rocket power to propel a heavier camera-laden lander to another world. But the introduction of new, more powerful rockets promised to change all that. Driven by the desire to find life on Mars, US interest had turned to the red planet, leaving project Venera to snatch the first Venusian vista.

With the USA's *Viking* landers already on their way to Mars to steal the glory of the first image of an alien landscape, the first Venera lander to carry a camera to Venus was dispatched in an attempt to clinch this prize. It touched down successfully on 22 October 1975 and with no time to waste, the protective lens caps were jettisoned and the first views of Venus were beamed up to the orbiter. The resulting black-and-white, 180-degree panoramas of this forbidden planet showed an overcast, flattish landscape, strewn with flat rocks. Sharp edges on the rocks

showed that erosion on the surface was much less intense than expected. *Venera 10* arrived five days later and returned an almost identical picture from a landing site several hundred kilometres south. The camera faced the ground, but the top corners of the picture gave a glimpse of the unforgiving, oppressive atmosphere that had caused so many problems. These were the first pictures returned from the surface of another world.

Planets are normally mapped before landers are sent, but the reverse happened in Venus's case because of the challenge of 'seeing' through its dense clouds. Crude radar mapping had already been attempted from Earth, but in December 1978 NASA's *Pioneer Venus Orbiter* reached Venus with a radar mapper and an altimeter to chart the planet's terrain. The *Pioneer Venus Multiprobe* followed a few days later. It comprised four atmospheric probes that descended through the night and day skies of the planet, returning data on temperature, pressure and chemical composition of the atmosphere. Over the next eight months, the *Pioneer Venus*

Orbiter charted 93 per cent of the surface from orbit, revealing the patterns of plains, basins and mountain chains.

We had to wait till 1982 for the next surface pictures, when *Veneras 13* and *14* managed to land in the southern hemisphere, northwest of a plain called Lavina Planitia. This time the pictures were colour. More flat terrain and flat rocks were visible, and an X-ray analyser was to establish the chemical composition of the rocks. But as fate would have it, a lens cap from *Venera 14*'s camera fell right where the analyser was to reach, preventing it from making contact with the surface. *Venera 14* touched down more than 20 years ago and nothing has attempted to land on Venus since.

Two more Venera radar orbiters and a couple of weather balloon missions helped improve our charts of Venus in the 1980s, but a new style of mission was in the pipeline. NASA's *Magellan* probe had had its budget halved before construction even began, so it had to be assembled from leftovers of earlier projects like Voyager. *Magellan* orbited Venus from pole to pole, taking it over every part of the planet as Venus rotated slowly beneath it. Between September 1992 and October 1994, Magellan systematically mapped 98 per cent of the surface at a resolution of up to 120 m (394 feet). The project was a huge

success, and the results transformed our understanding of Venus, its crust, and the geological processes that have shaped the planet. But they also threw up new mysteries, and solving these may well require more durable robots that can spend months on the planet's surface.

1. Launch of *Mariner 2* – the first successful interplanetary spacecraft bound for Venus.

2. View from *Venera 14* – snapped through a diamond window pain. Note the lens cap on the ground in front.

3 & 4. *Venera 9* took the first ever images of another planetary landscape in 1975.

5. Another view of the *Venera 14* landing site – note the sprung sampling arm that landed right where the lens cap had fallen.

6. A *Magellan* radar image of the surface showing three steep-sided, flat-topped volcanic domes – the largest is about 60 km (37 miles) across.

7. NASA's *Magellan* spacecraft before launch.

The Red Planet

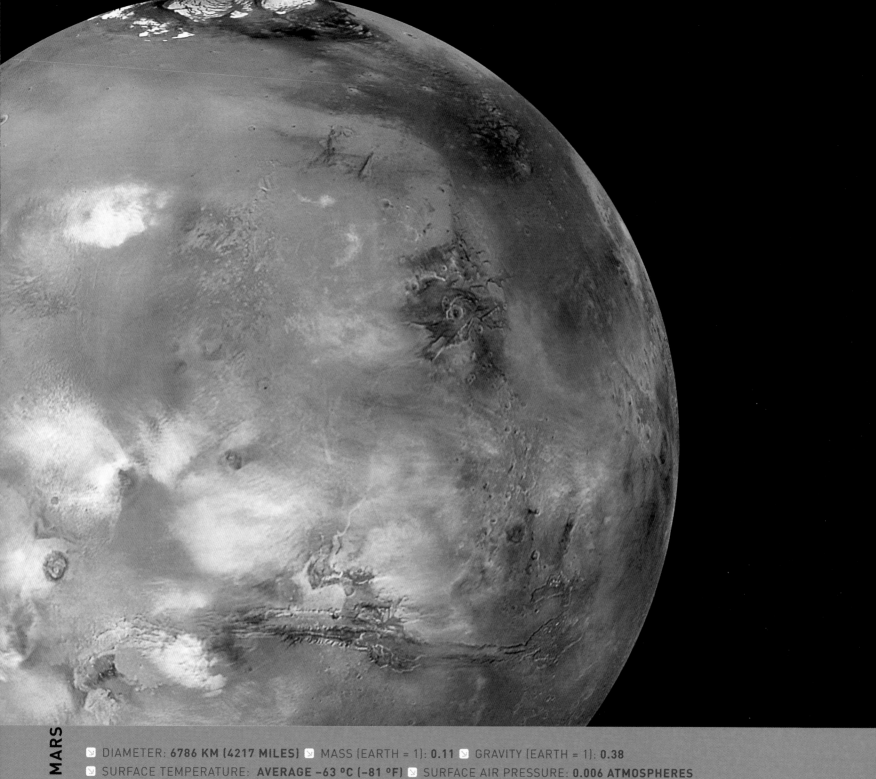

MARS

- DIAMETER: **6786 KM (4217 MILES)** • MASS (EARTH = 1): **0.11** • GRAVITY (EARTH = 1): **0.38**
- SURFACE TEMPERATURE: **AVERAGE –63 ºC (–81 ºF)** • SURFACE AIR PRESSURE: **0.006 ATMOSPHERES**
- DAYLENGTH (ROTATION PERIOD): **24.6 EARTH-HOURS** • YEAR: **687 EARTH-DAYS**
- DISTANCE FROM SUN: **228 MILLION KM (142 MILLION MILES)** • RINGS: **0** • MOONS: **2**

The Venus landing was only the first of many triumphs for *Pegasus* but, at the time, the pressure to succeed was so enormous that it was hard to appreciate our achievement. If, after all the money and resources that had gone into the project, we had fallen at the first hurdle, I doubt whether we would have had the option to continue. So it was with some excitement and a great deal of relief that everyone at Control started to turn their attention to the much more welcoming prospect of a landing on Mars.

Yvan and Zoë recovered well from the landing and the crew spent a further 10 days in Venusian orbit, taking pictures of its exotic atmosphere and measuring wind speeds and other weather phenomena. Geologists pored over data from the sensors that Yvan had left on the surface – they continued to function for almost 60 days, despite the extreme environmental conditions.

On 15 March our trajectory window opened, and at 08.32 universal time, Control gave the all clear for the engine burn that would launch *Pegasus* on its 62-day flight to Mars. For the support staff, this moment was as exciting as the landings. The craft's huge main engines opened up, and within 10 minutes I was watching the crew pressed into the backs of their seats by the acceleration. Within an hour *Pegasus* had reached its cruise speed of just under 240,000 km/h, and Tom's voice echoed around Control: '*Pegasus* has its wings wide open and is flying straight to Mars'. This was met with huge applause.

Over the next two months the sense of anticipation grew to fever pitch. Of all the planets, Mars is the one that truly grips the public's imagination. It has inspired centuries of science fiction, and robotic landers added further fuel to our curiosity about the prospects of life on Mars. I grew up with posters of the *Viking*, *Pathfinder*, *Spirit* and *Opportunity* landing sites on my bedroom walls, and I've lost count of the number of times I've looked through those ridiculous 3-D glasses at red-and-blue images of Mars, trying to imagine I was there. The prospect of finally conducting a human exploration of this legendary planet was tremendously exciting. However, no-one at Control was fooled by the benign appearance of Mars. We all knew that, despite recent successes, more than half the missions to the red planet had ended in failure. Even after conquering the hell that was Venus, we had no time to be complacent.

Alex Lloyd
Chief Scientist, *Pegasus* Mission

❚ Nina Sulman, Mission Scientist
Mission elapsed time: 75 days, 16 hours, 12 minutes

The chores of daily life on *Pegasus* leave little time for science. A ship this big needs a lot of attention – decades of space-station life have taught us that. But during this interplanetary cruise, life is far less regimented than when we were in Venusian orbit. Our tasks are presented in blocks, and the crew can arrange their whole day themselves. I find that a routine keeps me focused, and the days seem to rush past. I wake to an alarm every morning at 8.00. By 8.30 I am finished with breakfast, and I run a programme of experiments until 11.30. (Yesterday I made a thorough examination of the Venus atmosphere samples, and although I confirmed the presence of the peculiar chemicals observed from Earth, I couldn't see signs of Venusian life forms.) Then I spend an hour or so running on the treadmill and using expanders, which helps maintain my muscle strength in the low gravity. After clean up, if time permits, I have lunch with the rest of the crew and we talk about our day. I do science experiments for another four hours and then spend an hour on the exercise bicycle before dinner. After dinner I usually spend one to two hours preparing for the next day's work. By the time I've finished it's about 10.00 p.m., and I'll spend the next two hours looking out of the cupola windows or watching movies.

People often ask if there's a typical astronaut, but there really isn't – we're all individuals. Everyone has a different personality and a different skill mix. Tom's always preoccupied with our relationship with Control. He's often very serious – although he's got a dry sense of humour too. He's also quite a spiritual man and sees some kind of greater significance in this cosmos than I confess I do. But he's always been the one we all respect. Yvan is very focused on keeping *Pegasus* fully functioning. I never see him without a tool belt on or a head torch pointing into some shadowy corner, panels pulled back and wires out everywhere. I've known him for years, and I've noticed that the Venus landing has changed him a little, given him more confidence. One might say it has made him a bit cocky. Well, he earned it.

I think Zoë is more sociable than me – always anxious that everyone is getting along and that she's seen to be pulling her weight, doing all the chores. Before leaving Earth she sold her house and put all her possessions into storage. She decided that if she didn't have a home back on Earth to worry about, *Pegasus* would become her home. There is no doubt it has.

John is the most distant of all of us. He's often locked away in his sleep module discussing our health on a secure link to the flight surgeon at Control. I know he's responsible for our mental and physical health, but I think he could lighten up a bit sometimes. But it's these people that make this experience what it is, and I wouldn't swap any of them.

✳ Tom Kirby, Commander
85 days, 11 hours, 14 minutes

We've had so many questions e-mailed to us about *Pegasus* that I thought I'd take this opportunity to say a bit more about the spacecraft we call home. Well let's begin with some vital statistics. From tip to tip *Pegasus* is 1300 m in length – you could fit 12 football pitches along her. The whole structure weighs around 400 tonnes. The most obvious feature is the giant disc at one end. It is 400 m across and shields us from both the Sun's radiation and the planetary atmospheres we fly through when slowing down. It's made of carbon and beryllium and can withstand temperatures of more than 5000 ºC.

The disc also houses a powerful nuclear fusion reactor that serves as our main engine. A giant girder, or truss, keeps our living quarters a safe distance from it. The spherical tanks inside the shield contain our fuel – liquid hydrogen – which is pumped into the reactor core, heats up, and comes surging out of the main nozzles, pushing us in the opposite direction. It's a phenomenal engine – eight times as powerful as the *Saturn V* rockets that first took men to the Moon. In deep space it can accelerate *Pegasus* to a maximum speed of 280,000 km/h. We refill the hydrogen tanks from fuel dumps sent ahead of us into orbits around our destination planets and moons.

In the middle of *Pegasus* is the cluster of modules where we live. They store food, air and water and house the recycling systems that make our life away from Earth possible. Solar panels and radiators are mounted here as well, the former to help generate electrical power for some of our systems and the latter to remove excess heat from our living area. Would you believe that staying cold in space is harder than staying warm. At Venus the temperature in sunlight outside *Pegasus* was 195 ºC. It's about

50 ºC cooler now, but that's still hot. And all the machines and power systems inside the ship generate heat too. If *Pegasus* didn't have the radiators, the temperature in here would be intolerable.

Protruding from the habitation modules is our command centre – a kind of cockpit where we control *Pegasus* and manage the lander excursions. Directly beneath this is the cupola – our windowed observation area, built from five-layered, gold-tinted glass. It's bulletproof and could withstand a direct impact from a piece of debris the size of a golf ball travelling at more than 100 km per second.

The other obvious feature of this part of *Pegasus* is our gravity simulator. This consists of a set of pods on centrifugal arms that rotate around the ship at up to three revolutions per minute, pushing everything outwards with a force of about half Earth's gravity. We exercise, sleep and wash in this part of *Pegasus*. We try to spend at least 10 hours a day here for the sake of our bones and muscles, which are strengthened by exposure to even this level of artificial gravity. You enter the pods by floating into the rotating module at the centre of the ship and then letting the force pull you in. You've got to hang on for the last bit and climb down a ladder or you could fall!

At the other end of the ship is a Y-shaped structure – another important power source. It contains radioactive material, which is why it is separated from the habitation area by a 300-metre truss. Inside each arm of the Y, heat from radioactive plutonium turns a fluid into hot gas, which in turn drives turbines to generate power. The fluid is then cooled and pumped around the system again.

It's a big, complex ship. I suspect only Yvan knows all the systems thoroughly, but we all understand a bit of it well, and together that allows us to look after our interplanetary home. And just in case we miss something, there's a team of 300 people back on Earth keeping a constant eye on *Pegasus*'s systems via radio links.

The Red Menace

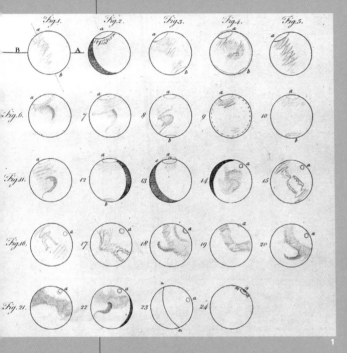

For thousands of years Mars has been a source of both fascination and fear. With its blood-red colour and its strange, looping passage through the constellations, this wandering red star became a symbol of chaos and war to ancient civilizations. The Greeks and Romans named it after their gods of war – Ares and Mars, respectively – and they understood its orbit and learnt to predict its appearance. But little else was discovered about Mars until the invention of the telescope.

Being cloud-free and close enough to see in detail, Mars was an obvious target for the first telescopes. The Dutch astronomer Christian Huygens is credited with the first scientific observations of Mars, made in 1659. He discovered and sketched the dark patch known today as Syrtis Major, and kicked off three centuries of speculation about the planet's inhabitants. A century of telescope improvements allowed William Herschel to observe seasonal changes around the poles in the 1780s. The discovery that Mars had seasons led Herschel to conclude that 'its inhabitants probably enjoy a situation in many respects similar to our own'.

In the 1850s an Italian priest called Pietro Secchi reported the presence of dark lines – which he called *canali*, or 'channels' – on the surface of Mars. Within 20 years, the Italian astronomer Giovanni Schiaparelli had mapped and named the channels. Encouraged by the mistranslation of *canali* into the English 'canal', the American astronomer Percival Lowell convinced himself and others of the presence of an elaborate irrigation system designed to take water from the Martian poles to the more arid equator. The myth of the Martian canals was born, and the public lapped it up. Even rocket pioneer Robert Goddard was romanced by Lowell's theory; the liquid-fuelled rockets that he designed in 1926 were inspired in part by his ambition to send a spacecraft to the canal-covered world.

The existence of Martians was a 'scientific fact'. France's Guzman Prize – an award of 100,000 francs for the first person to make contact with extraterrestrials – even excluded contact with Martians as too easy an option. Science-fiction writers were cashing in too. Inspired by Lowell, H.G. Wells penned *The War of the Worlds*, and Edgar Rice Burroughs described the Martian adventures of his hero John Carter, on a planet that resembled southern Arizona.

Like Herschel a hundred years before him, Lowell was convinced that Mars had an Earth-like climate, which he described as 'no less comfortable than the south of England'. But not everyone was convinced. In 1907 Alfred Russel-Wallace came closer to the truth, describing the planet as a perpetually frozen, dead place as forbidding as the Moon.

Gradually, a more realistic view of Mars developed, but that didn't stop the *New York Times* running a front-page story in 1921 when radio-pioneer William Marconi announced that he'd picked up wireless messages from a civilization on Mars.

And in 1938, Orson Wells's radio dramatization of *The War of the Worlds* – in which a realistic news broadcast described a Martian invasion of Earth – sent a million Americans into a panic. The US public was jittery, and the military reacted by drawing up detailed maps of the Martian canals. Despite the lack of credible evidence for intelligent life on Mars, such maps were still being published by the Engineers Corps as late as 1964.

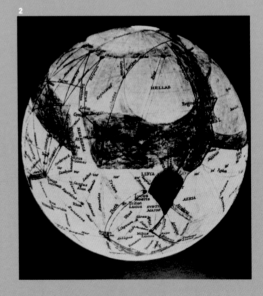

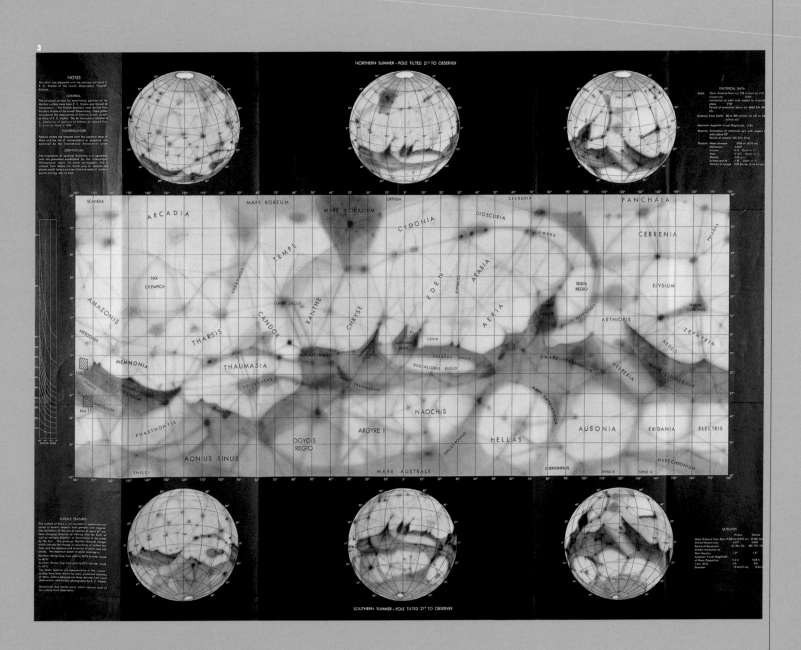

1. Herschel's drawings of Mars published in 1784.

2. A map of Mars drawn by Percival Lowell in 1907.

3. US air force map of Mars from 1962 still showing the 'canals' promoted by Lowell.

✚ John Pearson, Flight Medic
98 days, 17 hours, 5 minutes

We're now only 19 days from Mars orbit. The planet is pulling us in at 64 km a second, and it looks clearer with every moment. For the three of us that are going to the surface, it is getting increasingly difficult to concentrate on mundane tasks. I keep trying to find an excuse to sneak to the cupola for a look. The 15 days we spend on that surface are going to be the most precious of my entire life.

But, for the medic in me, there are worries. With no magnetic field or ozone layer, Mars doesn't provide much of a shield against radiation from space. Every solar storm that flies by scours the surface and if one occurs whilst we're out on the surface I can't guarantee our safety. The three of us going down there have already upped our radiation-protection diets – more fresh fruit and antioxidants. I can see Tom is anxious too. But there's nothing I can say to assure him that this is going to be safe, and that's a very helpless feeling.

💬 Zoë Lessard, Mission Scientist
119 days, 7 hours, 32 minutes

We reached Mars orbit 48 hours ago, and after a series of orbital trim manoeuvres, we're now in a 4.5-hour equatorial orbit 2500 km above the planet. Unlike cloudy Venus, the feeling of scale here is immense. That vast expanse of rust-red desert stretching out below makes me catch my breath every time I visit the viewing chamber. In a single glance I could see from the giant volcanoes of the Tharsis Bulge, dominated by the outline of mighty Olympus Mons, to the vast canyon of Valles Marineris, and farther to the dark lands of western Sinus Meridiani.

At the most northern extent of our orbit I can see the white tongues of carbon dioxide and water ice fields that spiral out from the north pole. I can't pretend I don't want to be one of those to go down there, but the flight plan calls for just three people to land, and exobiology – the search for Martian life – is deemed more important than geology. So Nina is going instead of me. Anyhow, John's an accomplished geologist as well as a medic I guess, so

the expedition is in safe hands. And I am here in orbit and I'm not expecting any clouds to get in the way, so I've got a ringside seat. After all I suppose history will only remember Tom Kirby as the first man to walk on Mars.

☒ Yvan Grigorev, Flight Engineer
121 days, 11 hours, 32 minutes

I have just completed our first refuelling spacewalk, and what a fantastic experience! At first, as I exited *Pegasus*, I panicked a bit. I got an attack of vertigo looking down at Mars, and I found myself gripping the handrail a bit too hard. I haven't felt like that on an EVA (extravehicular activity) for a very long time, and it took me a bit by surprise. I must have spent 300 hours doing spacewalks above Earth during the construction of *Pegasus*, but somehow, floating here above this dry, almost cloudless world, felt very exposed – like swimming in deep ocean water.

For the refuelling manoeuvre I needed to venture further than normal from the main body of *Pegasus*. The fuel tanks are vast: 3 km long – which is more than twice the length of *Pegasus* – and 500 m wide. They hold almost 600 million cubic metres of gas each. Once we were docked, we pumped the hydrogen into our cryogenic tanks, where the gas is compressed and cooled into liquid.

For at least 90 per cent of the time out there, you're totally focused on the task. You get wrapped up in what you're doing, you get the job done, and you move on. But at one point I was waiting for some backup advice from Tom, and there was time to just look. There's just 4 mm of toughened glass between my face and the extraordinary vista. The colours were incredible – from the lightest, brightest ochre to the darkest rust-red, with wisps of white cloud painted across this global canvas. The highlight was when Tom suddenly said, 'You've got to stop what you're doing and look at what's beneath you right now'. I tightened off the clamp I was using and swung my body upwards so I could take a look down. And there, passing beneath us, was a giant hunk of rock more than 20 km wide. It was Phobos – Mars's innermost and largest moon, a captured asteroid that orbits the planet faster than Mars turns. I could even make out the shadow it cast on the planet below – a tiny eclipse scurrying across the deep red Martian surface. Wow.

On an EVA in Martian orbit Grigorev helps refuel *Pegasus*.

Earth Invades Mars

Although no Martians have ever been found, if they are aware of our attempts to explore their home planet they probably have a very dim view of our technology. Over the last 40 years we have sent an armada of robots to Mars – over half of

Earth. Suitable launch windows for Mars only come up every 26 months when the red planet is nearest and in the early days both superpowers attempted to launch at each opportunity.

Neither had much success until 1964, when NASA's *Mariner 4* spacecraft made it all the way to Mars, passing within 1000 km of the surface. Its 21 blurred black-and-white shots revealed a disappointingly barren, cratered world, not dissimilar in appearance to our own dead moon with an ultra-thin carbon-dioxide atmosphere, no magnetic field, no canals and certainly no

NASA's *Mariner 9* spent about a year studying the atmosphere and photographing the surface. They discovered dried-up river channels, giant Martian volcanoes and a vast valley around its equator. The new Mars that had emerged from these pictures was a world where life might have once lived and still lurk. The next phase of missions had a new goal – to hunt for Martians.

Following the cancellation of the Soviet Mars programme after a series of very public mission failures it was NASA's Viking programme that resumed exploration of

which have either missed the planet all together, burnt up in the atmosphere or smashed into the surface. Our exploration of Mars has been a story of determination rather than easy triumph. And things didn't start well. In October of 1960, only three short years after the first ever satellite *Sputnik*, two Soviet spacecraft were on the launch pad ready to head for Mars. But they didn't get much further, failing to reach parking orbits around

evidence of life. Interest in Mars temporarily dried up and as the next launch window slipped by neither superpower made an attempt for the red planet.

1971 was an auspicious year for Mars. The planet would pass closer to Earth than it had done in 60,000 years and seven Soviet and US spacecraft were being readied to go. Although two Soviet landers did reach the surface they didn't send back any useful data, but their orbiters and

Mars in 1976. Four spacecraft conducted a five-year close-up study of the planet, but it was the biology experiments carried by the landers that attracted most public interest. Initial results seemed to be positive – but closer inspection lead to the conclusion that it was chemical and not biological reactions that the experiment had detected. The lack of any organic material in the soil was not encouraging either. Twelve years and six launch windows passed

Overleaf: *Ares* glides down to the surface of Mars – crossing the vast volcano Olympus Mons. The colossal white-and-blue striped paraglider canopy needed to fly in this low density atmosphere would cover a football field.

without a mission being sent to Mars as funds for space exploration were diverted into the Space Shuttle Programme in America and space station endeavours in Russia.

Then in 1988 a Russian mission named Phobos flew two new Soviet spacecraft to study the planet's atmosphere and drop landers onto a Martian moon. The mission was planned to kick-start a programme designed to land humans on Mars by 2015. Again both were bitter disappointments. The failures continued – in 1993 NASA's *Mars Observer* mission went quiet just as it

had reached Mars. Russia's ambitious international *Mars 96* mission ended up scattered across the Andes after a launch failure.

Undeterred, NASA launched their sophisticated *Mars Global Surveyor* (*MGS*) orbiter and *Pathfinder* lander mission at the next opportunity in 1996. There had been no completely successful mission to Mars for 20 years and a success was desperately needed. *Pathfinder* bounced

down onto the surface of a large outflow channel called Ares Valles, 800 km (497 miles) south-east of the *Viking 1* site in July 1997, and dispatched a little six-wheeled rover called *Sojourner*. Over the next three months it scrutinized the rocks and

boulders around the landing site, finding them to be chemically similar to igneous rocks on Earth; probably eroded off nearby lava flows and transported there by flood-waters one thousand times more powerful than the Amazon.

Back in orbit, after an initial period of risky aerobraking, to place it in the right orbit, *Mars Global Surveyor* had begun to send back unprecedented views of the Martian landscape. Breathtaking views of

layered canyon walls, fresh channels etched into cliff lines and the tangled paths of giant dust devils poured in each day. *MGS* continues to provide unprecedented views of Mars's surface, changing theories and throwing up new questions, which have made us realize how little we really understand this planet.

With the success of *Pathfinder* and *Global Surveyor* in the late 1990s, and invigorated by the debate over life in a Martian meteorite, four more NASA missions were heading for Mars in 1999. But their *Polar Lander*, *Climate Orbiter*, and *Deep Space 2* microprobes all failed. Depressed but undeterred, the quest to better understand the red planet spurred NASA back to Mars and in 2001 their *Mars Odyssey* craft arrived in orbit and started a mission to analyse the chemistry of the Martian surface and map the water content of the subsurface. It's already discovered huge amounts of water ice in the north polar caps – hinting that as much as 50 per cent of the upper metre of soil might be water ice.

1. The first image from Mars by NASA's *Viking 1*.
2. Over 20 years later, *Pathfinder* and the *Sojourner* rover.
3. The first close-up image of Mars returned by *Mariner 4* in 1964.
4. A dust storm rages in Syria Planum – *MGS*.

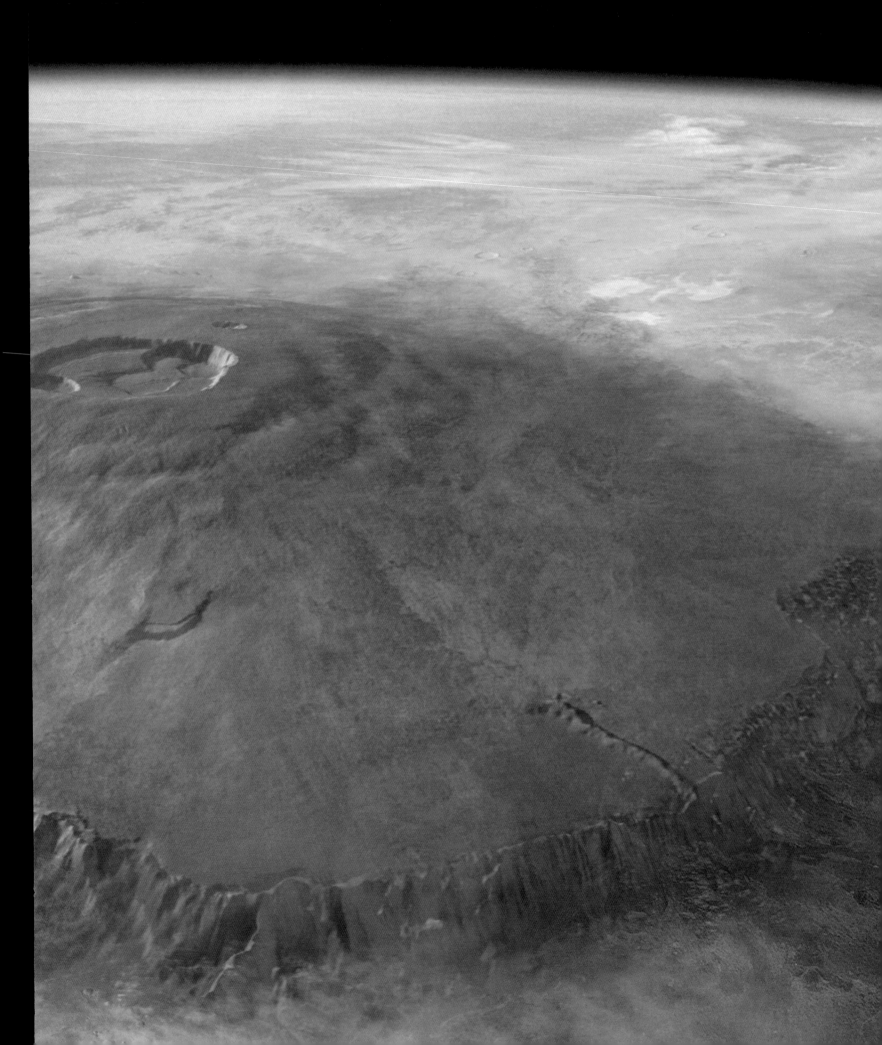

Twenty-first-century Mars

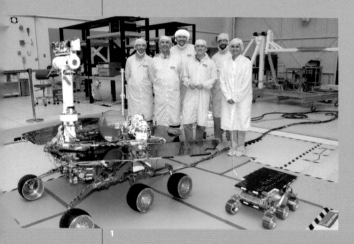

Just 56 million km (35 million miles) separated Earth from Mars in August 2003. The planets were even closer than they'd been in 1971 and this time there were more than just two space agencies who were hurling probes towards Mars. Japan, Europe and America had all mounted robotic expeditions – taking advantage of the shorter distance and less fuel needed to reach our neighbour.

From the ashes of Russia's *Mars 96* came the European Space Agency's *Mars Express* – an orbiter and a British lander called *Beagle 2* in celebration of Charles Darwin's epic voyage. Despite being right on course, *Beagle 2* failed to contact home after leaving its mothership *Mars Express*, on 19 December 2003. Repeated attempts to reach it using NASA's *Mars Odyssey*,

radio telescopes around the world and two weeks later *Mars Express* itself all failed, and the mission was classed as lost in February 2004.

By then two NASA Exploration Rovers called *Spirit* and *Opportunity* had made textbook landings on opposite sides of the planet. These sophisticated rovers were designed to look for more evidence that Mars had been wet in the past, rather than attempt to detect the direct presence of past or present life. On 3 January 2004

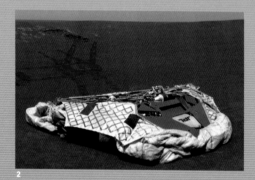

Spirit had landed in Gusev Crater – returning pictures revealing a far flatter, less rubble-strewn landscape than any previously seen on Mars. It was perfect terrain for the wheeled rover to explore and after initial software teething troubles a thorough exploration of the location began. Gusev Crater was a gigantic outwash channel – where scientists

suspected a vast river delta had dumped sediment into a lake several billion years ago. Sure enough strewn across the location were carbonate minerals that could have formed in the presence of water. *Spirit* carried a low-powered microscope to examine details in the rocks and found spherical particles thought to have been rounded by water.

Meanwhile ESA's *Mars Express*, now in a polar orbit, was also on the lookout for evidence of water – sending back images of the terrain of the highest resolution ever achieved in the history of our exploration of the planets. The European spacecraft collaborated with NASA's *Spirit* rover – making simultaneous observations of the Martian atmosphere from above and below. Sensors on board the spacecraft also confirmed the presence of water ice at the south pole – only implied by previous spacecraft observations.

Mars Express carries a powerful ground-penetrating radar instrument that can detect water up to five kilometres beneath the crust, five thousand times deeper into the crust than *Mars Odyssey* can see. Mission scientists plan to look at the whole story of water on Mars – from ice reserves beneath the surface to water vapour lost into space from high in the upper atmosphere.

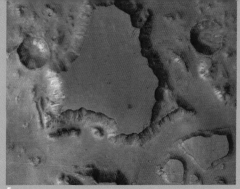

Around the other side of Mars the twin rover, *Opportunity*, was busy exploring a plain called Meridiani Planum. It had landed on a flat smooth plain, at the highest altitude ever attempted by NASA, coming to rest inside a small crater. The terrain was quite unlike anything else seen at a Mars landing site and left the spacecraft team 'astonished'. In the first pictures sent back geologists could see a light-coloured layered rock outcrop in the walls of the crater – the first bedrock ever witnessed close up on Mars. Closer inspection revealed that they contained sulphur salts that could only have been formed in water. Other minerals found at the site could also be formed in the presence of water. There were even imprints of ripples in the rocks as if they'd been formed under a current of water on the bed of a lake or a salty sea. Quite how long the water was around for, or even whether it was a body of water exposed to the air or trapped beneath a sheet of ice is impossible to say.

On Earth such a watery environment would be a prime place to hunt for fossils. Confirming whether or not there are fossils in the rocks at Meridiani Planum might have to wait for a future rover mission to investigate. If life had evolved in environments like this it is likely that it was just microbial and *Opportunity* doesn't carry the right instruments to find such tiny life forms.

But the geological wonderland that *Opportunity's* landing site turned out to be means that this location is a prime one to send a sample return mission to in the future. The question of whether or not there is or was once life on Mars will continue to carry us back to the planet in search of answers for decades to come.

1. Those who assembled it gather around one of NASA's 2004 rovers. A flight spare of the earlier *Sojourner* rover is next to it for comparison.
2. The rover *Opportunity* looks back to photograph its landing cocoon.
3. A view from *Spirit* – its heat shield can be seen on the left as a tiny bright dot.
4. The small outcrop of 'sedimentary' rock discovered by *Opportunity*.
5. A mesa on Mars captured with the high resolution camera on board ESA's *Mars Express*.
6. Mars - its main features are labelled in black. The white labels show the landing sites of robotic spacecraft. M2 - *Mars2*, M3 - *Mars3*, V1 - *Viking 1*, V2 - *Viking 2*, MPF - *Mars Pathfinder*, MER-A - the *Spirit* rover, MER-B - the *Opportunity* rover.

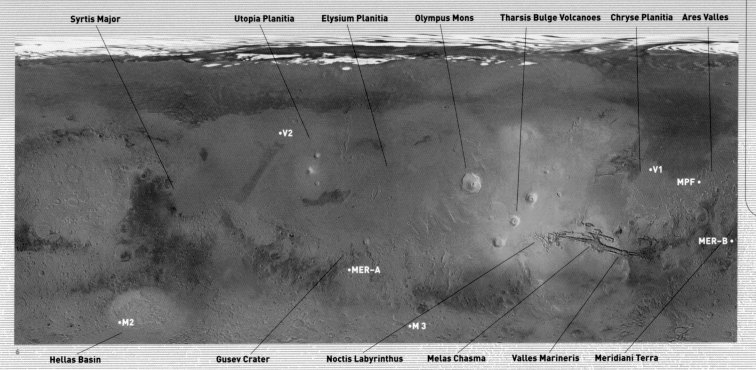

Syrtis Major   Utopia Planitia   Elysium Planitia   Olympus Mons   Tharsis Bulge Volcanoes   Chryse Planitia   Ares Valles

•V2

•V1

MPF •

MER-B •

•MER-A

•M2

•M 3

Hellas Basin   Gusev Crater   Noctis Labyrinthus   Melas Chasma   Valles Marineris   Meridiani Terra

Ares approaches the landing site at Melas Chasma on the southern edge of Valles Marineris.

ℹ Nina Sulman, Mission Scientist
123 days, 15 hours, 22 minutes

This is my first diary entry from the surface of another world, and so far Mars has been everything we expected and more. Even as I stare out across the red boulder fields of our landing site, I just can't believe I'm really here. 'Marineris Base' is our camp for the next 15 days, or 'sols', as we call them here (a Martian sol is 37 minutes longer than a day on Earth). It's 19.15 local time and the Sun is already low in the sky to the west – the dust-laden red atmosphere showing tinges of blue as the light passes through the thickest part. It's the first proper sunset I've seen in four months. It will soon be dark, and the temperatures outside will drop to 90 degrees below freezing. But we won't have trouble sleeping after a day like today.

The descent in *Ares* was a rough ride. As it entered the atmosphere, the lander began to heat up and shake violently. Arcs of plasma tore clumps off the heat shield and licked around the windows. John let out a whoop of excitement as the first Gs kicked in, but I started to feel waves of nausea. We were coming in over Arcadia Planitia, a smooth, featureless desert in the northern hemisphere. Yvan was watching from *Pegasus* – he said we looked like a huge meteor streaking eastwards across the red terrain. Between bouts of sickness I could make out the northwest lava fields of Olympus Mons. Travelling at 13,000 km/h, it took less than a minute to cross the volcano's vast crater summit. The sky was turning from black to indigo, and below us water vapour appeared in the thin air, forming wave clouds in the turbulent winds around the summit.

Passing the last big volcano (Pavonis Mons) and still 28 km high, *Ares* released a drogue parachute, and a tug of gravity kicked in as we decelerated abruptly. The roughness of the ride eased as we went into a wide swing, and the capsule began to spin gently below the drogue. The dark skies of space we'd become so accustomed to were now growing pink. Some 20 km lower, the heat shield beneath us was catapulted away. Simultaneously, the drogue was jettisoned and a vast blue-and-white paragliding canopy unfurled to steer us over Noctis Labyrinthus – a chaotic, jumbled terrain we'd first seen from Martian orbit a few days before. Even with the giant parachute open, *Ares* was falling too fast through the thin air for a safe landing, and a blast from the rocket engine was needed to decelerate us further. We had enough fuel to manoeuvre around the edge of Melas Chasma, clear the hills behind, and pick a smooth and level landing site. Tom piloted as John read out figures from the descent computer, providing a running commentary on our progress. I just sat in the middle watching the fuel gauges and tank pressures.

The prongs on the legs touched the surface and a light blinked on. We were in contact with Mars! 'Marineris Base here. *Ares* has landed.' Tom announced.

The journey had taken only 40 minutes, but by the time Tom and John were ready to enter the airlock and step outside, it was already after midday local time. I glanced at the live camera feeds. They were all working – one on the lander leg, and two on each spacesuit. Reception was good. I could see Tom opening the hatch and inching out feet first, while John stood inside, talking him out.

I watched Tom's helmet camera as he swung around to face the view and stepped onto the surface. 'With this first step I tread lightly on a new world, not to claim it for this generation but to preserve it for our children.' He later told me he'd adapted these words from an old North American Indian saying. It seemed appropriate, given his roots and our determination to protect these pristine environments.

It was shortly after this that Mars gave us a real red-planet welcome. Tom was describing the texture of the surface when I noticed the dust around his legs starting to shift. At first I thought it was an illusion, but the dirt really was moving. The dust shifted in bursts at first, but soon there was a continuous stream of it, pulled by some invisible force towards a point about 100 m in front of him. And to Tom's left and right, the same thing was happening. Tom noticed it very quickly and abandoned his running commentary of the spacewalk. For a few moments, we were at a loss to understand what was going on.

Glancing upwards – higher than Tom could see from his helmet – I saw a large column of dust gathering above. It hadn't touched

the ground and I was sure Tom couldn't see it, so I told him on the radio that a dust devil was forming above him. Tom turned around to face *Ares* and instructed John, still in the doorway, to seal the hatch. He looked back to face the column of dust. Martian dust devils are much bigger than their Earthly cousins. This one was about 50 m across and must have been several kilometres high. Sparks of static crackled around it. There wasn't much Tom could do as he stood his ground, bracing himself. I lost sight of him from the window and looked at his helmet camera feed. All I could see was a red fog and flashes of electrical discharge which interfered with our radio communications.

It seemed to go on for minutes, but as the column of dust moved past and off to the east, Tom was revealed again, standing in the same spot. His brilliant white spacesuit was now red, and I could

A dust devil towers over Kirby within minutes of him stepping onto the surface.

Kirby – after his Martian welcome.

no longer make out his face through his dust-blasted visor. Once we'd established that he was all right and done some suit checks, I began to think of the hours of cleaning that lay ahead of us if we were to ready his suit for the following day. Fine dust can work its way deep into the joints and seals in a spacesuit. Left uncleaned, the suit's moving parts can seize up and fail, with potentially disastrous consequences. Tom's suit would have to be disassembled, thoroughly cleaned and relubricated, and then put back together again.

We decided to abandon the spacewalk and bring Tom back inside. Just as I'd suspected, the fine Martian dust was everywhere. It had somehow got through to his skin in places, and by the time we'd given his suit a thorough check, it was all over our skin too. It seems to have permeated our pores and has spread a musty smell through *Ares*.

The Sun has now set on our first Martian day. Out of the window, the rust-red landscape has turned black, the temperature has dropped to –60 °C, and the wind has shifted direction as the cold night air is pulled eastwards towards the dayside of the planet. Bright stars are visible. I can see Jupiter high above the western horizon – easily the brightest object in the sky. Our next destination is still more than 900 million km away, but it's intensely bright and seems to be beckoning us across the solar system.

- - - - - - - -

◾ Nina Sulman, Mission Scientist
124 days, 16 hours, 3 minutes (Mars sol 2)

The excitement of being on Mars made sleep all but impossible last night. I was still awake at 3.30 a.m. local time, when I watched Earth rise in the east. There was our home planet – a brilliant blue star. This is Mars and that is Earth. How weird.

By 4.30 a.m. there was an eerie twilight in the eastern sky, though sunrise was still about 90 minutes away. The dust-laden atmosphere scatters the sunlight over the horizon long before the shrunken sun comes up. By the time the Sun had risen properly, all three of us were ready to descend the ladder. As soon as the seal was broken on the hatch, I heard the wind whipping through it. Tom went first, guided by John, and John followed, guided by Tom. I brought up the rear, guided by John, and clumsily made my Martian debut – on my knees – as I crawled backwards out of the hatch.

I dropped down the last two rungs in one slow-motion hop, enjoying my first real taste of one-third gravity, and took an overexcited jump off the footpad and onto the surface, revelling in the billowing clouds of talc-like red dust that rose from my boots and hung in the air. I quickly found it easier to bound than to walk and perfected my own energy-efficient Martian toe-skip. The dust squeaked like snow as my boots compressed it. Tom keeps calling it sand, but sand needs silicon and there's no silicon on Mars. All this dust has formed from the weathering of Mars's basalt crust.

I looked across at John, who was larking about like a small child, leaping into the air and enjoying the new sensation of falling slowly. I reached down to pick up a rock and noticed something glint in the red dust that I'd disturbed. Staring back at me was a glassy patch of green mineral. I picked it up and, following a field geologist's reflex, tried to blow dust off the sample – which of course made my visor steam up. Sheepishly, I resorted to the brush on my tool belt. The mineral turned out to be olivine, which meant that the landing site had been cold and dry for a very long time. Dating it could help us find out exactly when Mars became so dry. I radioed the find to Zoë; she sounded jealous.

Our primary task for the second sol was to erect our base camp – a specially fabricated, UV-proof shelter that provides a temporary store for samples and equipment. We also had to inflate a hydrogen balloon that carries a radio transmitter; unpack and assemble our three-person rover; and unpack a number of robotic explorers. Weather permitting, the robots should outstay our visit by several months – the severe temperature changes that occur from day to night are usually what kills them.

Temperatures are well below freezing here, and there were patches of frost on the ground until about mid morning. I could feel the cold through my boots and had to turn up the heating elements in the soles. The suits are very good. Our own metabolic heat is enough to keep the whole body warm. In fact I sometimes get too warm and the suit has to work to keep me cool, even though it's –20 °C outside. We'll be spending up to 10 hours a day inside these life-supporting second skins, and generally they are comfortable to work in. The most annoying thing is when you get an itch on your face. If I can scratch it on the helmet microphone, that's fine – but I can't always reach, and then the itch torments me until I'm distracted by something else.

As the day passed, the ground warmed up and I readjusted the heating elements in my boots again. My suit's sensors revealed that the outside temperature was now 20 °C warmer around my ankles than around my head. The air pressure on the surface of Mars is barely a hundredth of that on Earth, and the ultra-thin air traps very little of the warmth emanating from the sun-warmed ground. How different this planet is from Venus! The Sun climbed higher as the day passed, but the weaker, reddish light felt constantly like it was a late evening on Earth. Robbed of our instinctive means of judging the time, we had to keep an eye on the clocks inside our helmets to stay on schedule. Those clocks will rule our lives for the rest of our stay on Mars.

Pearson and Kirby (behind) – the first men on Mars.

Melas Base on the southern side of Valles Marineris.

Nina Sulman, Mission Scientist
127 days, 12 hours, 15 minutes (Mars sol 5)

As usual we were up before the Sun, preparing for another packed day. After just five hours rest, the stiffness in my muscles brought on by yesterday's busy day had completely evaporated. Control thinks this is because our hearts are more efficient at pumping blood in the low gravity.

Today's plan was to drive 15 km in the rover to the lip of the Valles Marineris canyon. Once there, we were to launch an ambitious sample-return robot called *Charlie* (after Charles Darwin), which is designed to visit the valley floor. The air pressure at the base of Valles Marineris could be high enough for subsurface water to exist as a liquid for a few days in the year. *Charlie* has a ground-penetrating radar to home in on possible patches of water, and a drill to bore for a sample.

Kirby and Sulman assemble a robotic explorer called Charlie (after Charles Darwin)

As we loaded up and boarded the rover, there was still a thick frost on the rocks around camp, and ice crystals in the high atmosphere threw bright rings around the freshly risen Sun. Solid H_2O was all around us. I took this as a good omen – perhaps *Charlie* would find some in a liquid state. Our route had been carefully planned from satellite images and advance robotic exploration. Navigation is impossible with a compass on Mars – the planet has a very weak

magnetic field, a relic of times when it was more geologically active. So instead we used a laser range-finder to measure the distance to prominent landmarks, which allowed us to triangulate our position on a digital map. Fortunately for us, the whole of Mars was mapped in great detail by NASA's *Global Surveyor* spacecraft, and ESA's *Mars Express*. (Ironically, since there are no oceans to

Chapter 2

hide surface details, we've got more complete maps of this red world than of our own.)

It took just over an hour to reach our destination, the digital map alerting us to our proximity to the edge of the giant canyon long before we glimpsed it for real. We parked a safe distance from the cliff and walked cautiously towards it. It's hard to find words to describe the view we found before us. We were facing north, looking across the widest part of this continent-sized chasm. A few shreds of mist hovered over the canyon, but between them the air was crystal clear. The 7-kilometre deep valley tumbled away from us in a succession of landslides to the vast valley floor, where giant ridges and grooves stretched into the distance. The drop was almost the same height as Mount Everest!

We were just returning to the rover to ready *Charlie* when the solar flare alarm kicked off in our earpieces. We knew too well what this meant and we didn't hang around to discuss the next move. All back on board the rover, we headed for base camp at full speed.

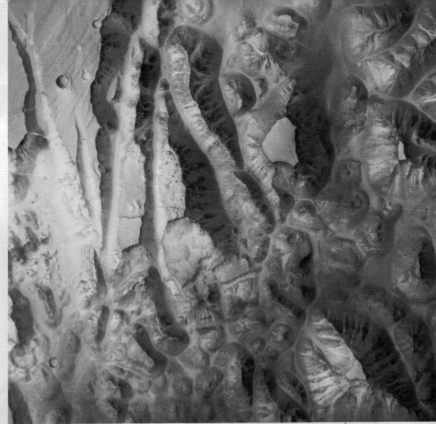

A portion of Valles Marineris from *Mars Express*.

Tools of the Martian trade.

During a solar flare (or solar storm), the Sun hurls out a lethal cocktail of radiation and charged particles, which hurtle across the solar system and start to impact Mars within 45 minutes of erupting. And this one, as Yvan told us en route, was an S-class – S standing for severe. Prolonged exposure to such space weather could prove fatal. Our rover offered some protection, but we really needed to ride out this storm inside *Ares*. With luck, we'd be back before the nasty stuff arrived.

As I write this, we've been back in *Ares* for more than 12 hours. The forecast is that gamma radiation will remain high for a couple of days, which will really muck up our schedule. Radiation is a strange, invisible threat. Our bodies are not designed to sense such danger and so, despite our knowledge of its presence, it's hard to really fear it until it's too late. Outside *Ares*, Mars looks just the same as yesterday – and yet our sensors show it is tens of times more dangerous.

Nina Sulman, Mission Scientist
132 days, 15 hours, 18 minutes (Mars sol 10)

The solar flare was a major setback, and after careful consultation with Control we dropped a series of experiments in favour of resuming *Charlie*'s mission to search for water on the floor of Valles Marineris. This morning – five days after our first visit – we drove back to the cliff, and the view was no less stunning. We inflated the hydrogen balloon that was to carry *Charlie* down, placed him securely in a gondola slung below the balloon, and sent him on his way. My helmet microphones picked up the sound of the motors in the gondola's wings as it began to steer into the prevailing easterly winds, revving furiously in the thin air. *Charlie* headed for the floor of the canyon, about four hours flight away. We watched the view from his camera on a laptop on the Mars rover.

I could make out aeons of volcanic strata in the steep cliff walls, each layer of rock a testimony to Mars's turbulent geological past. Strewn across the vista beneath were eroded desert features: canyons showing the wear and tear of millions of years of sandblasting; worn rocks and boulders; fluted canyon walls; and eroded butts and mesas.

Riding some lucky downdrafts of air, *Charlie* reached the canyon floor in about three and a half hours, faster than expected, and started to circle over the ground as he looked for signs of

The Red Planet

**Overleaf: From the southern edge of Valles Marineris Kirby and Sulman watch an approaching canyon storm.
The canyon floor is some seven vertical kilometres beneath them.**

subsurface water. He quickly identified three promising sites and landed to investigate them. It was on the third of *Charlie*'s boreholes that we noticed a rise in conductivity in the soil, suggesting liquid – perhaps salt water. We cheered and whooped with excitement. *Charlie* now seemed to be drinking. I knew he was designed to do this, but I was incredulous that he'd found liquid water on Mars. With no time to waste before the exposed water vaporized, *Charlie* siphoned a sample into a pressurized, insulated collection vessel and begun to sink a fourth hole.

Kirby peers over the edge of Valles Marineris.

In our excitement we missed the first of Zoë's warnings. From orbit she could see a dust storm at the western end of the canyon. It had just emerged from the nightside of Mars, and she estimated it to be four hours away at its present speed. These canyon dust storms are caused by the difference in temperature between the night and day ends of the canyon, and we knew they were a common occurrence. Still, we had hoped to avoid them. I estimated the sample return would take three hours, and Tom gave me the go-ahead to start the ascent. The darkening storm front was now 600 km west and beyond our view, and we knew the consequences of hanging around too long.

We had to allow at least 50 minutes for the drive back to base camp before visibility deteriorated. But as the sample canister rose it was blown eastwards ahead of the storm front and away from the camp. We had to board the rover and set off in pursuit. As we chased the balloon east, the western horizon behind us

began to darken visibly and flashed with lightning. *Charlie* finally cleared the cliff tops, and I steered him out of the canyon and onto the plateau before rapidly deflating the balloon to bring him down.

With the sample safely stowed, we turned and raced back towards *Ares*, heading west and directly into the menacing storm front. Light levels fell and visibility plummeted as we drew near to camp, and John had to follow our outbound tyre tracks for the last two kilometres. There was no time to make camp secure. We rushed up the ladder and closed the hatch as the wind began to howl and the dust started swirling around. Wind speeds were pushing 100 km/h and the ground was trembling. But all I was concerned about was that the water sample was safely on board with us.

The *Ares* crew monitor robot *Charlie's* progress – hunting for water 7 km below.

★ Tom Kirby, Commander
136 days, 10 hours and 5 minutes

Marineris Base is now 300 km beneath us. Although happy to see Yvan and Zoë again and my *Pegasus* home, I miss the Martian surface – it was every bit as captivating and awe-inspiring as I had hoped. The red planet didn't exactly give us an easy ride – with the giant dust devil welcome, the solar flare, and the three-day dust storm – but at no time did I feel like we didn't belong. In those final days on the surface, after clearing up from the storm, it truly felt like home. Out of direct communication range with Earth on the last evening, we played softball at base camp for 30 minutes – losing four or five balls in the desert before abandoning the game and standing together to enjoy the sunset. High whips of cirrus

water-vapour clouds caught the last rays of sunlight, and for a moment it felt like a new Earth.

The next morning, *Ares'* main engine lifted us back into orbit for a rendezvous with *Pegasus*. Beneath us the engine was generating its own dust storm, but we climbed smoothly away from the landing site. Together with our robot helpers we managed to collect a huge amount of data while we were down there.

At 4000 m up, *Ares* rolled over and I caught a last glimpse of our base camp, the many footprints and tyre tracks clearly visible in the soft dust. I wondered how long it would be before the constant rain of fine dust obliterated them. After a few more storms, no-one would be able to tell we'd been there at all. Back on our home planet there has long been talk of 'terraforming' this neighbouring world – chemically reengineering its atmosphere to turn Mars into a second Earth. Our success could well revitalize this debate. But I think to attempt such vandalism would be to destroy what we'd come to explore. We all believe the planet should remain a pure wilderness used only for scientific research.

It was strange at first to be back on *Pegasus*, with its bright, pristine, weightless rooms and its different smell. Yvan pointed out that we not only smelt of Mars but looked like Martians. The red dust is ingrained into our skin, and no amount of washing seems to get rid of it. But I am strangely proud to be red. In days to come, when Mars shrinks to a tiny red star in the cupola window, the sight of it will bring back memories of base camp and that unforgettable view of Earth in the pre-dawn sky. And I'll remember a time when Mars was my home.

Clearing up after the dust storm at Melas camp.

Life on Mars

Well into the twentieth century it was a commonly held belief that Mars was awash with water, vegetation and even advanced civilizations. All sorts of fanciful ideas abounded, including suggestions that Mars was home to giant beavers with thick fur for the cold and huge eyes to see in the dim sunlight. But from the 1960s closer robotic encounters with the planet buried these ideas.

At the time of the Viking landers experiments it was thought that life could not exist without sunlight, water and oxygen. Ironically just as the landers were scraping Mars topsoil looking for microbes that fitted this description, life was found hiding inside rocks lying in the Dry Valleys of Antarctica. Within a year more of these exceptional bugs were discovered around smoking volcanic vents on the floor of the Pacific Ocean – thriving in super-heated, pitch black, waters. The discoveries prompted a new definition of life and the conditions needed to sustain it.

This new definition came too late for the Vikings, but the landers did make at least one important contribution to the hunt for life. They sampled the Martian air and recorded its chemical composition. This told us that a handful of peculiar meteorites found on Earth were probably from Mars – air bubbles trapped inside them carried the same chemical signatures as the Viking landers had found on Mars.

Why go all the way to Mars to look for life when we could look in our collecting cabinets here on Earth? In August 1996 a NASA team at Stanford University announced that they'd found 'compelling but not conclusive evidence of primitive life' in a 3600 million year old, potato-sized Martian meteorite called ALH84001, which had been found in Antarctica in 1984. The evidence included organic compounds and 'fossils' of what appeared to be microscopic Martian worms.

But after years of debate the controversy boiled down to whether or not the evidence was really from Mars or simply contamination from Earth. The only way to settle the argument was to go and look for life on the surface of Mars itself.

But what would we go to look for and where should we look? ALH84001 had also brought into sharp focus the problems of agreeing about how to even recognize extraterrestrial life. Life reproduces, mutates and selects. These are not characteristics that robotic missions could easily detect. So the next best thing to search for are the ingredients we think life needs to survive; energy, liquid water and oxygen. On Mars the limiting ingredient is water.

The temperatures and pressures are so low that water can exist only as a gas or a solid, yet the landscape suggests that things weren't always like this.

Geologists have amassed a great deal of evidence that Mars really was once awash with water. There are sinuous channels crossing desert plains, gullies running down cliffs, and dried up lake and sea beds. There are even minerals in the rocks that require liquid water to have formed. If estimates are right, Mars has enough water to flood its surface with an ocean 500 m deep. But how long did the water last? For life to flourish water would have to stay on the surface for millennia, forming oceans, rivers and rain. But some scientists think it made only a fleeting

Chapter 2

appearance, perhaps released by volcanic eruptions or asteroid impacts melting underground ice and briefly flooding the surface, before the water seeped back into the ground and refroze. For three decades this issue has underpinned much of the debate about life on Mars.

So in the images beamed back from our Mars-orbiting satellites and our surface exploration rovers we continue to hunt for proof of past lakes and places where hydrothermal waters reached the surface. These are the sites that future missions will visit with the hope of finding fossilized evidence that life at least used to live on Mars.

The question of whether Mars is still alive today might be easier to answer than we think. In March 2004 it was announced that methane had been detected in the Martian atmosphere. The unstable gas had first been spotted from telescopes on Earth and was eventually confirmed by ESA's *Mars Express* orbiter. Methane should be unstable in the Martian atmosphere – hanging around for just 100 years at best. So its discovery implied that there was something still producing it.

We know of only two sources for methane. Volcanoes and microbes. The former would mean that Mars is still geologically active, an exciting new discovery in itself, but one that might also mean that there are areas where heat from underground might turn the ice deposits to liquid water; a vital ingredient for life. But if the methane is not volcanic then it's likely to be coming from bugs called methanogens. Here on Earth these primitive life forms require very little

oxygen and produce methane as a by-product. Ironically if the *Beagle 2* lander had been successful it could have answered this question about the source of the methane. If there was ever a reason to rebuild and re-fly a spacecraft, this is it.

1. A sequence of pictures showing the *Viking 2* lander sampling Martian soil.

2. The controversial carbonate minerals that some attribute to Martian microbes, found in a meteorite from Mars in 1996.

3. Water ice frost coating the rocks around the *Viking 2* landing site.

4. Channels apparently carved by water are abundant on Mars.

5. Water ice clouds at the sunrise terminator of Mars taken by the Hubble Space Telescope.

6. Gullies in the walls of impact craters perhaps formed by the release of groundwater.

7. Water ice (left) and carbon dioxide ice (middle) deposits seen by ESA's *Mars Express* spacecraft at the south pole. The right-hand image is what the human eye would see.

Through the Sun's Corona

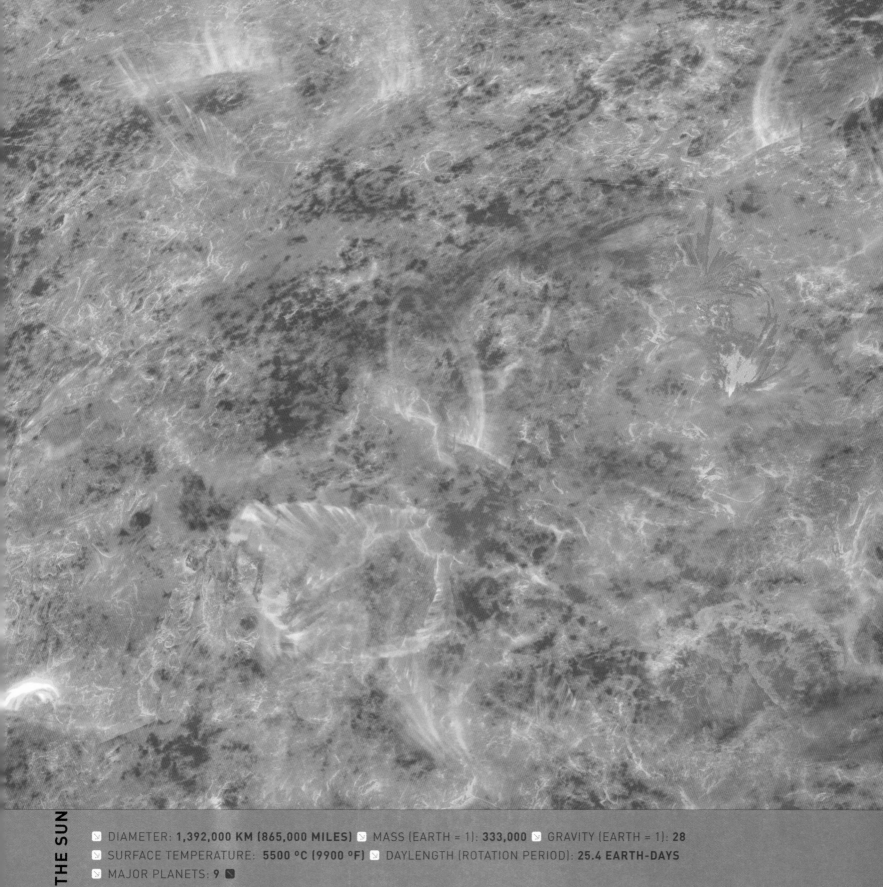

THE SUN

DIAMETER: **1,392,000 KM (865,000 MILES)** · MASS (EARTH = 1): **333,000** · GRAVITY (EARTH = 1): **28**

SURFACE TEMPERATURE: **5500 ºC (9900 ºF)** · DAYLENGTH (ROTATION PERIOD): **25.4 EARTH-DAYS**

MAJOR PLANETS: **9**

There were many firsts on this mission, and after Mars and Venus, another was drawing to a close: the first human space flight through the inner solar system. *Pegasus* had already generated enough new data to keep the science community busy for decades. A scientific conference was quickly convened after the successful discovery of a tiny sample of Martian water from the water table 1.4 m beneath the floor of Valles Marineris. The sample was just 0.05 ml of liquid – about a hundredth of a teaspoonful. It was hardly enough to even call a drop, yet it ignited so much interest on Earth that it might as well have been the corpse of a Martian. It was deemed too precious to drop off when *Pegasus* passed Earth orbit three months after leaving Mars. Nina carried out limited analysis, but we were all itching to get hold of it, praying that no contamination ruined it. Of course, as the mission continued, *Pegasus*'s growing cargo of extraterrestrial samples had become more than priceless, and there were arguments about whether the grand tour should be cancelled and the payload returned before anything happened to it. In the end we decided to continue – things were going too well to abandon the mission.

After turning its back on Mars, *Pegasus* headed straight for the Sun. Jupiter, the next destination, was orbiting on the opposite side of the solar system, 1120 million km away – twice as far as the craft had already travelled since leaving Earth nearly five months before. We didn't have enough fuel to make the journey unassisted in the time allowed, so we attempted something that had never been tried before: a so-called 'Sun burn'. Mid-course corrections were made to ensure that, as the spacecraft fell towards the Sun, it would pass within 8 million km of the solar surface. *Pegasus* would then make another engine burn at the point of closest approach. Harnessing the Sun's immense gravity, which is 28 times that of Earth, would accelerate *Pegasus* to a staggering 300 km per second (a hundredth of the speed of light) and propel her across the last 800 million km to Jupiter in just 208 days.

At least, that was the theory. The biggest hole in our plan was that a rogue CME, or coronal mass ejection, might erupt from the Sun's surface. If *Pegasus* had been caught by one of these gargantuan sun storms it would have finished the mission. As it turned out, the flight past the Sun was yet another spectacular success for *Pegasus*, but it pushed the crew to the limit and ultimately led to tragedy.

Alex Lloyd
Chief Scientist, *Pegasus* Mission

Zoë Lessard, Mission Scientist
Mission elapsed time: 147 days, 10 hours, 12 minutes

Our windows are filled once more with that black void of interplanetary travel, dominated by the glare of the Sun. I've never been able to get used to the Sun shining in a black sky. No matter how long I spent in training above Earth's atmosphere, those black skies of day always seemed unreal.

It's hard to really appreciate the Sun's colossal size. We can never look directly at it, and for most of the journey so far it's been possible to blot it out with a thumb held at arm's length. It's easy to forget that behind that thumb is 99.8 per cent of all the mass in the solar system, packed into a star some 109 times wider than Earth. I sometimes try to picture its vast size by thinking how long it took us to orbit Earth from horizon to horizon and then multiplying that time by more than 100.

But the truly terrifying thing about the Sun isn't its size or mass but the overwhelming amount of energy it pumps out. To say that it is humbling is perhaps the understatement of our mission. Conditions inside this stellar furnace mock the harshest extremes of our entire voyage. At its core, temperatures hover around 15 million °C, and the tremendous weight of the overlying layers exerts a force 340 billion times greater than Earth's air pressure. These astronomical conditions make atomic nuclei fuse together, turning 5 million tonnes of matter into pure energy each second. Stars also release some nasty things along with the energy. At such high temperatures, dangerous X-rays and gamma rays are generated, and waves of harmful, charged particles wash out of the surface.

So why are delicate little biological structures like us flying towards it? It's not just to pick up speed to reach Jupiter. Flying through and sampling the Sun's outer atmosphere – the corona – will provide important data about one of the last unexplored regions of the solar system. For research, the most important period will be the two days either side of our closest approach. It will be a busy time, and I guess this will help keep our minds off the danger. Of course, we'll become the first humans to travel this close to a star. And since we're all ultimately made of stardust, I see it as a sort of homecoming!

John Pearson, Flight Medic
150 days, 8 hours, 20 minutes

- -

We might have left Mars behind, but my mind is still there, back in those unforgettable, dusty red plains. For 15 days we had a whole world to ourselves. I'm still clinging to my memories of the Martian surface, fearful of letting the image fade, but I know I've just got to appreciate and savour each astonishing vision before letting it go and accepting that it will be superposed with another. Adjusting once again to the confinement of interplanetary space flight has not been easy, but throwing myself back into work as we prepare for the brush with the Sun has helped.

The scientists among us are excited about visiting the 'last unexplored part of the solar system'. But I can't help thinking *Pegasus* should be renamed *Icarus*. My main preoccupation on this mission is with ionizing radiation – the waves of subatomic, charged particles that wash invisibly through the ship. Each tiny particle that passes through us knocks electrons off our atoms, ionizing them. The electrons shoot off and ionize other atoms too. And the ions that are created can start chemical reactions that damage DNA and lead to radiation sickness or cancer.

Some of this radiation comes from distant exploding stars, but most is from the Sun – and we are heading straight for it. Our trajectory will take us right through its outer atmosphere; there's nowhere more lethal in the entire voyage. Inside *Pegasus* we will do everything we can to shield ourselves from the radiation. On

Still 50 days away from closest approach to the Sun and the glare is unbearable.

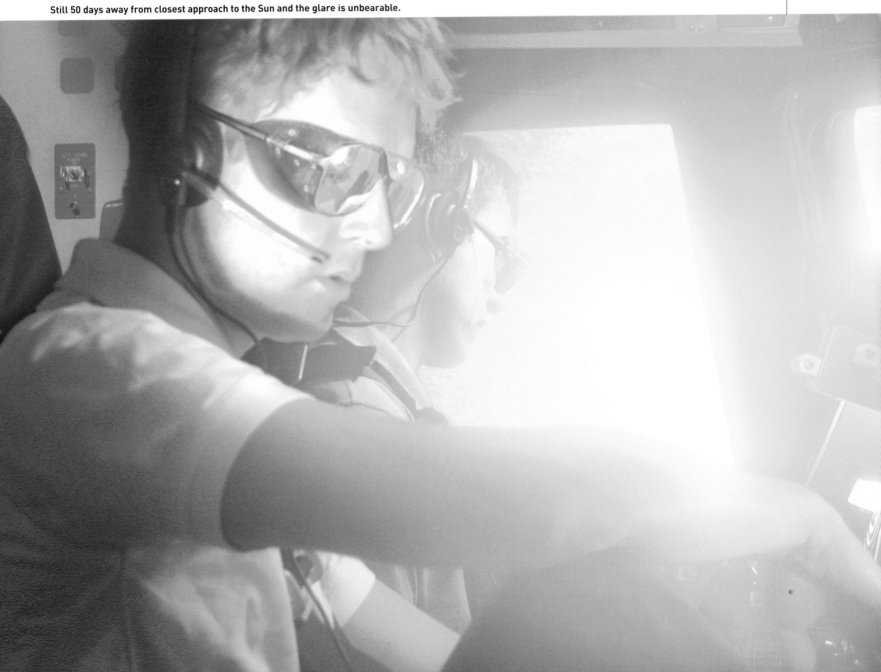

the surface of Earth we are cocooned by the planet's atmosphere and magnetic field. Here on *Pegasus* we will try and simulate the same things, using our stores of drinking water as a shield around the living quarters, and generating a temporary magnetic field around the ship during our closest approach. At least the ionizing radiation will kill any bugs that have got into the water.

We are passing the Sun during a period called the solar minimum, which is when the Sun's 11-year cycle of activity (the solar cycle) is at its quietest. But even during this quiescent time, it will be vital to keep *Pegasus*'s main shield constantly facing the Sun. The front of the shield will roast at more than 3500 °C, but our living quarters will stay in the relative safety of the shield's shadow. We will also have to power down a lot of the nonessential systems so that we can pump maximum power to our magnetic shield. This should deflect most of the high-energy particles we'll encounter at our closest approach.

In preparation for the Sun burn, we're upping our dietary intake of antioxidants (such as vitamins C and A) to mop up any dangerous 'free radicals' that might be created by radiation passing though us. And should the worst happen, we have an experimental medicine that can help damaged cells repair themselves (or in extreme cases execute themselves). It's not fully proven, and it's only to be used if something goes badly wrong.

Anyway, Earth is still clearly visible, and that makes me feel that we're not far from help, should a stray solar flare come our way. If disaster struck in the next few days, we could abort the mission and place ourselves on a two-month trajectory to Earth.

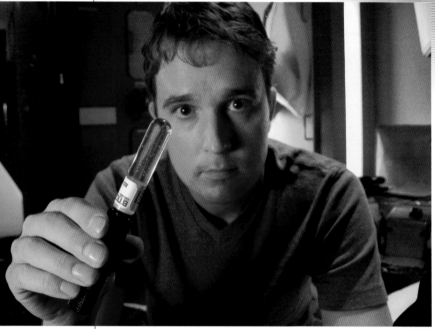

Pearson holds up one of the radiation detectors.

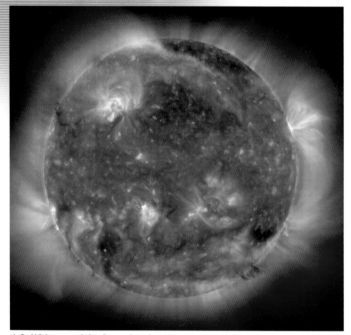

A SoHO image of the Sun taken in extreme ultraviolet light.

Some radiation will still leak through though. It's monitored around the ship by 'bubble tubes' – glass capsules of gel in which bubbles form when charged particles pass through them. We are also wearing personal dosimeters that change colour when exposed to the radiation. Green is good. We're all in the green at the moment, and things should stay that way. Strangely, we've all noticed that our eyes also act as radiation detectors: occasional tiny flashes of light happen when cosmic rays (rogue particles from distant exploding stars) hurtle through the retina or optic nerve.

**▮ Nina Sulman, Mission Scientist
177 days, 16 hours, 14 minutes**

For ages during the planning of our mission, statisticians and scientists haggled over the relative risks of trying to return our Mars and Venus samples to Earth as we crossed the inner solar system, versus hanging on to them for the rest of our mission. In

Chapter 3

the end they settled on a bit of both, and so in mid-June, passing Earth orbit, we jettisoned a small pod of samples. But the tiny fraction of Martian water we'd retrieved was deemed too valuable to return in this way. I was charged with making a thorough examination of it on board, before the solar flyby.

But after three weeks of patient scrutiny, and after checking and rechecking all my results and discussing them with Alex, I had to concede that there was no trace of life in the sample. The salty water contained organic compounds all right – and at levels that would guarantee it to be teeming with microbes if it was from

Pearson films Sulman as she breaks the seal on the Martian water sample inside the clean box.

From the moment I seized the sample canister on the Martian plains, I'd hardly let it out of my sight. I had to wait until a quickly convened science conference had collectively decided how to examine the water before I was allowed to touch it. That didn't stop me looking at the sample through its toughened glass vial and wondering what secrets it might hold. The anticipation haunted my sleep and stalked my waking hours. I could think of little else that month as we fell silently towards the Sun.

Earth. But this drop of Martian water held nothing that even hinted at life. I had to accept, reluctantly, that I was probably the only living thing that had been near this water. Of course, this does not prove there is no life on Mars or that there never was life on Mars – it just means it's not in this minute droplet of water.

From Godhead to Nuclear Furnace

PLANISPHÆRIVM
Sive
VNIVERSI TO·
EX HYPO·
COPERNI
PLANO

COPERNICANVM
Systema
TIVS CREATI
THESI
CANA IN
EXHIBITVM

People have been watching the Sun cross the sky each day since the dawn of humanity, so it is not surprising that early civilizations and even scientists fell into the trap of thinking the Sun orbits us. As a bringer of light and warmth, the Sun was also an obvious celestial object to deify. As soon as people could build, edifices were constructed in honour of it. The first standing stones, burial mounds, shrines and temples were dedicated to the celebration and worship of our nearest star. But the idea that the Sun was a star, and any concept of how it generated all that heat and light, were millennia away. Breakthroughs in unravelling the

mysteries of the Sun were slow to arrive, held back at first by technology and later by the Catholic Church.

In Europe the birth of solar science had to wait until Greek times. Both Philolaus (c. 480–405 BC) and Aristarchus (c. 310–230 BC) toyed with the idea of a heliocentric (sun-centred) solar system, but their work was lost in the great fire in Alexandria, and in any case Greek philosophy remained wedded to Aristotle's certainty that the Sun orbits Earth. It was a system that

also suited Christianity, and as the Church's influence on society grew, natural philosophers felt compelled to make their observations fit the established view, inventing increasingly complicated models to explain their observations. In 1543 the idea of a heliocentric solar system resurfaced in a book by the Polish astronomer Nicolaus Copernicus. His *De Revolutionibus Coelestium* (*On the Revolutions of the Celestial Sphere*) represented a lifetime's work, but because it was so controversial it had to be published after his death, with a caveat explaining that it was only a mathematical device and not to be taken literally.

The Earth-centred view prevailed for another 50 years until the Italian astronomer Galileo spoke out, publicly backing the Copernican system. The decision was to cost him his freedom. The Catholic Church placed him under house arrest for what they saw as an attack on the sacred, unchallengeable words of Aristotle and the Bible. In the grounds of his home prison, Galileo continued his astronomical work and became one of the first people to observe the Sun's surface through a telescope – a dangerous endeavour that robbed him of his sight in later life. Sunspots were first recorded by the Chinese in 28 BC, but Galileo is credited with the first observations of them in the West. Convinced they were fixed to the surface of the Sun, he mapped their movements around the disc and calculated that the Sun rotates in 27 days; the actual rotation period is 25.4 days. But the Church wouldn't

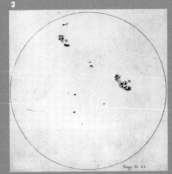

hear of such blemishes on God's perfect sphere – they preferred the explanation put forward by Jesuit astronomer Christoph Scheiner, who claimed the dark patches were silhouettes of planets passing in front of the Sun.

The spots remained a mystery for centuries. In 1781, astronomer William Herschel decided they were gaps in the clouds through which the solid surface was visible. Although Herschel made great discoveries in other areas of astronomy – he calculated that the stars were much further away than previously thought, for instance – he believed until the end of his life that the Sun's brilliant clouds hid a cool and pleasant world that might be inhabited.

After the Enlightenment, the Church's grip on science slowly weakened. In fact, it was a churchman who provided the next piece in the jigsaw puzzle. Theories linking the Sun with the stars had been around for decades, but it was not until 1862 that the Italian priest Angelo Secchi discovered that

they are made of the same stuff. Secchi was experimenting with the new science of spectroscopy, and he recorded the chemical composition of the Sun with a spectroscope placed on the roof of St Ignatius Church in Rome. Later that year he recorded the spectra of several stars and found they were the same. The missing link had been found. Not only was the Earth not a special world at the centre of the Universe, but neither was the Sun.

Refinements in spectroscopy over the following half-century led to the discovery of helium in the Sun, a finding that helped solve the mystery of how the Sun 'burned'. The Sun clearly couldn't be powered by ordinary combustion, which would burn through its fuel much too quickly. There had to be a mechanism that kept the Sun shining for billions of years, but such a mechanism would not be found by observation. The answer came from the new science of nuclear physics in the

twentieth century.

The helium, physicists reasoned, could be formed by fusion of hydrogen nuclei in the Sun's core. This nuclear fusion reaction would release vast amounts of energy without burning quickly through the Sun's mass. As Einstein's famous equation ($E = mc^2$) revealed, the amount of energy (E) locked up in matter equals the mass (m) multiplied by the speed of light squared (c^2). In other words, a tiny amount of mass can be turned into a vast amount of energy, and this is what happens during nuclear fusion. In 1926, the British astronomer Arthur Stanley Eddington calculated that a supply of fuel of Earth's mass could keep the Sun going for 25,000 years. Given the Sun's stupendous size, it should have no trouble shining for billions of years.

By the mid-1950s, new observations and further theoretical work on the nuclear-fusion model led to predictions about what the interior of the Sun should be like. It would take another half-century for advances in technology to confirm this theoretical work and unveil what lay beneath the Sun's dazzling surface.

1. A depiction of the new Copernican architecture of the solar system drawn around 1543.
2. An armillary sphere – built much later to explain the Copernican system.
3. Galileo's sketches of sunspots made in 1613.
4. Stonehenge, first built around 2900 BC to mark the midsummer sunrise.

⊠ **Yvan Grigorev, Flight Engineer**
193 days, 7 hours, 17 minutes

Ten days from the Sun, and we are making daily small engine burns to fine-tune our approach. Tomorrow we will turn the main shield towards the Sun, and increasingly frequent tweaks of the positional thrusters will ensure that it stays that way as we sweep past.

In another week the magnetoshield will be switched on. We were allowed one test before we started our present course, and all seemed fine – but it must perform flawlessly to keep us alive. The list of repairs to be done grows so fast that I can barely keep up with it. These daily problems are much more onerous than I'd thought they would be. Radiation is the biggest nightmare – the high-energy particles streak through computer chips, silently rewriting software. We've taken to routinely reinstalling critical systems from backups each day, rather than bothering to trace and repair errors that creep into the programming. I hope I won't have to spend the next five years patching up and making do like this.

◉ **Zoë Lessard, Mission Scientist**
194 days, 6 hours, 2 minutes

While preparations for the Sun burn go on in earnest, I'm leading the programme of observations of Mercury. Back on Earth, along with all the other geologists, I made the case that we should try and include Mercury in our grand tour. I must confess I am now relieved our arguments did not succeed. The workload on *Pegasus* is so great at the moment that I am glad to do no more than observe the planet through a telescope. Mercury is less than half the diameter of Earth and currently lies some 50 million km away, but its sun-facing side is perfectly illuminated, and the observations are proving fruitful. Only one spacecraft, *Mariner 10*, has visited Mercury, making two flybys of the planet in 1974. As fate would have it, the side we can see is the same one that *Mariner* surveyed, but at least we can look to see if anything has changed.

Seen through our telescope, Mercury's heavily cratered surface looks just like the far side of the Moon. The planet is a fossilized relic of the catastrophic bombardment that went on in the solar

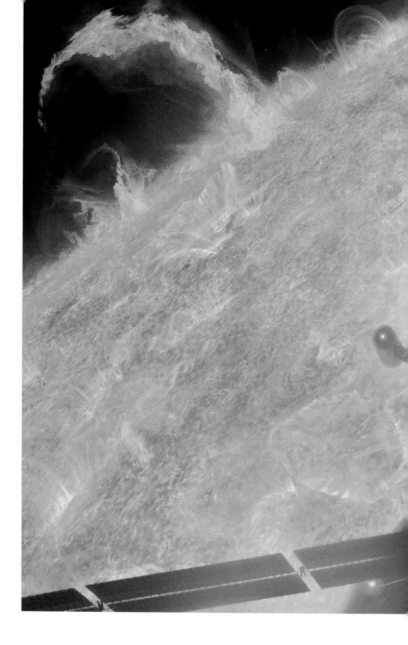

system's early history. In some places we can see what look like bright rays surrounding craters – these are streaks of fresh debris thrown out by more recent impacts. Smooth flat plains flood the regions between craters and a single large basin called Caloris stretches 1300 km across Mercury's face resembling the lunar mare (seas). Unlike the Moon, Mercury has a series of deep cliffs up to 3 km high that run for hundreds of kilometres. They probably formed early in Mercury's history when the planet was cooling and its crust was shrinking.

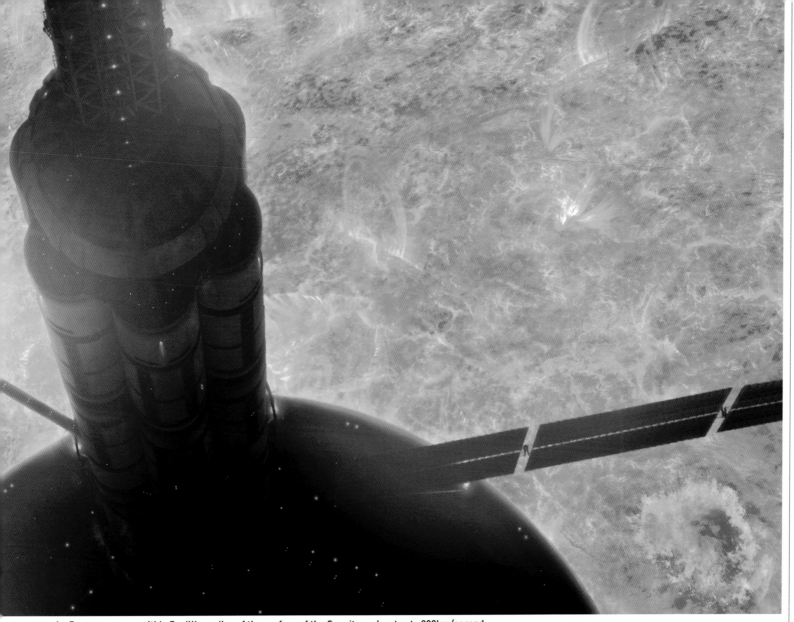

As *Pegasus* passes within 5 million miles of the surface of the Sun, it accelerates to 300km/second.

✦ Tom Kirby, Commander
202 days, 5 hours, 15 minutes

We're now just 10 hours from our closest approach to the Sun. At this proximity it is very different from the friendly object that crosses our skies on Earth. Its restless fury contrasts to its steady presence we feel on Earth. All the shutters are down on *Pegasus*, but we can see the Sun's seething surface all too clearly on the monitors. I can't help feeling that it's almost disrespectful to stray this close to a star. It is not a very welcoming place, even from millions of miles away.

This sense of unease is not helped by the change in routine that the flyby has forced on us. We confine ourselves to the heavily fortified command centre, and cameras on the far side of the main shield relay real-time images of the Sun. Automatic systems should pick up any dangerous solar activity, but I've also put Zoë and Nina on 'Sun watch' – they take alternate shifts to observe the screens in the darkened command module. For the last day, much of the ship has been in darkness. Every spare watt of power is being poured into the magnetoshield. Cameras outside *Pegasus* are picking up a faint aurora at the magnetoshield's poles – the rippling curtains of light are caused by charged particles striking a weak 'atmosphere' that leaks out of the ship and clings to the habitation modules.

This morning we crossed the orbit that our solar weather satellites occupy. An armada of these satellites is spread right around the Sun to watch for impending solar flares and coronal ejections. But as of today, we are closer to the Sun than any of them. If anything happens on this side of the Sun, we will know about it first.

Beneath the Glare

For years our endeavours to look deep inside the Sun were thwarted by its blinding surface. In the end it was seismology – the study of vibrations caused by sunquakes – that unlocked the secrets of its inner structure. The Sun, like most stars, is a colossal ball of hydrogen gas. The higher the temperature of a gas, the faster vibrations will travel through it. Density, composition, and even the motion

of the Sun's interior could all be deduced by studying sunquakes.

At the heart of the star, and where all its energy comes from, is a core 560,000 km (350,000 miles) in diameter. The phenomenal weight of the Sun's outer layers compresses the gas in the core with a force some 340 billion times greater than atmospheric pressure on Earth. Under these extremes the gas is squeezed to a density 10 times that of gold or lead, and

temperatures reach 15 million °C (27 million °F). Atoms are broken into their constituent parts, which slam into each other with enough force to overcome their electrical repulsions. The result is nuclear fusion: hydrogen nuclei fuse to form helium nuclei. In doing so, the hydrogen nuclei lose just under 1 per cent of their mass, and this mass is converted into radiation energy.

Trapped deep inside the Sun, the radiation cannot fly straight out into space. Instead it begins a very, very long journey to the surface. The energy is initially released in the form of gamma rays. These collide with hydrogen nuclei, and the energy re-emerges as X-rays. The X-rays eventually reach the layer outside the core: the radiative zone. This zone, which is cooler and far less dense, insulates the core and contains its ferocious heat. Even so, the radiation from the core takes up to 170,000 years to pass through the radiative zone, bouncing randomly from atom to atom as it inches its way up.

The next layer is cooler still, at about 2 million °C (3.6 million °F). This is called the convective zone, as the energy travels through this part of the Sun by convection. The lower layers of the convective zone absorb heat from the radiative zone and then rise in vast plumes of scorching gas, cooling as they near the top and then

sinking again – much like convection in a pan of boiling water. Energy moves up through the convective zone much faster than through the radiative zone. Travelling at 300 km/h (186 mph), it can cross the last 200,000 km (124,000 miles) to the surface in little more than a week. There is a thin interface layer (the tachocline) between the radiative and convective zones, and this is where the Sun's complex magnetic fields seem to be generated, by a kind of dynamo mechanism. The magnetic fields rise through the convective zone to the surface, forming sunspots and other features.

Of course, the sun has no solid surface, so its 'surface' is defined as the visible outer layer through which we cannot see – the photosphere. Temperatures are lower in the photosphere, and energy passing through it turns into ultraviolet radiation and visible sunlight before escaping into space. This energy may have taken 10 million years to travel from the core to the surface, but it crosses the 150 million km (93 million miles) to Earth in just eight minutes.

1. The internal structure of the Sun. At the centre is the core - the site of the Sun's nuclear reactions where hydrogen is fused into helium. Above the core is the radiative zone where energy liberated in the core is transmitted as heat and light. Above this is the conductive zone where hot gas rises to the surface to release energy before falling back to be reheated. On the surface or on the photosphere a large crop of sunspots is visible in the top right corner.

Mercury

Seen fleetingly by the ancients around dawn and dusk, tiny Mercury has always been difficult to observe because it orbits close to the Sun and is usually hidden in the glare. Between 1924 and 1929, the French astronomer Eugenios Antoniadi drew up charts of Mercury, and they remained the best maps of the planet until the late 1950s. Antoniadi believed, wrongly, that Mercury's orbit around the Sun was synchronous, which means one face is in permanent darkness and the other in permanent sunlight. In fact, Mercury turns very slowly: its day (the length of one complete rotation) lasts 58 Earth-days, taking up two-thirds of its year (one complete orbit of the Sun). Turning so slowly, and with little (or barely any) atmosphere to protect it, Mercury takes the full glare of the Sun, pushing the daytime temperature to 450 °C (840 °F). At night the ground temperature plummets to −180 °C (−290 °F).

Attempts to improve Mercury's maps by use of telescope and radar were superseded in 1974 by the arrival of *Mariner 10* – the only spacecraft that has visited Mercury. *Mariner 10* was cleverly slingshot around the sun to give it three flybys. It discovered a severely cratered world, with no magnetic field and a very tenuous atmosphere.

More recently, radar telescopes spotted what seemed to be water ice at Mercury's poles. Many astronomers were sceptical, but subsequent studies have confirmed the presence of large quantities of hydrogen – interpreted as H_2O – in permanently shaded parts of poles. Any ice on Mercury probably came from comets and has lain undisturbed for millions of years. It seems remarkable that, even this close to the Sun, frozen water and perhaps even simple forms of life could lurk.

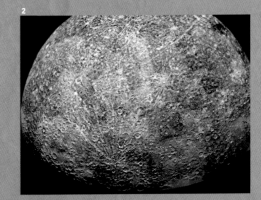

1. A view of Mercury from 125,000 miles away - taken by *Mariner 10*.
2. The southern hemisphere of Mercury taken by *Mariner 10*.
3. Mercury crosses the disc of the Sun in May 2003.
4. The edge of the largest basin on Mercury named 'Caloris' (Greek for hot). This depression is 1300 km (808 miles) across and is one of two regions that face the Sun when the planet makes its closest approach to our star. Temperatures in the basin at that time can reach 450°C (840°F).

Today I was the last to drift into the darkened command centre for our final Sun burn simulation. It was lit up like a Christmas tree inside, with all the monitors glowing in different colours. It was reassuring to remember that, 150 million kilometres away at Control, dozens of people were staring intently at the same images, and behind them were a thousand more technicians, and beyond them billions of souls all wishing us well. I feel reassured knowing that.

✴ Tom Kirby, Commander
202 days, 18 hours, 15 minutes

The worst is over. We passed perihelion (closest approach) more than three hours ago, and at last we're heading away from the Sun. It was a rough ride, and we're not out of the danger zone yet. We've been awake for nearly 35 hours, and the others have just bedded down. I'm staying up to record our voyage past the Sun before my memories of this extraordinary day are dulled by sleep.

Communication with Earth was deteriorating well before perihelion because of radio noise from the Sun, but we received an e-mail from Control with just an hour to go. It said we were 'good for the burn' and ended with the line 'Good luck to all of you'. Suddenly it felt very real. We were hurtling towards the eastern limb of the Sun at just under a million km/h. I looked around the command centre at the faces of my friends and then turned back to the looming face of the Sun on the monitors. It felt like a big, unstoppable tidal wave was surging towards us and we were about to ride it.

Yvan was glued to the data screens throughout the encounter, but the rest of us were fixated by the feeds from our sun-facing cameras. As we neared perihelion, the burns from the altitude thrusters became continuous, keeping our main shield directly in line with the Sun as *Pegasus* swung round. The temperature on the Sun-facing side was just over 3550 °C, but behind the shield it was a comfortable 20 °C. The real temperature around *Pegasus*, here in the fringes of the solar corona, topped two million degrees, but because there were so few atoms around us here in the Sun's outer atmosphere there was no proper heat. Nevertheless, we were sweating buckets because the air conditioning had been turned down to release power for the magnetoshield.

Even Yvan began watching the Sun as our final approach to this inferno-world loomed. Everyone was aware that the tiniest flicker of gas from its surface would snuff out our mission if it hit us. I found

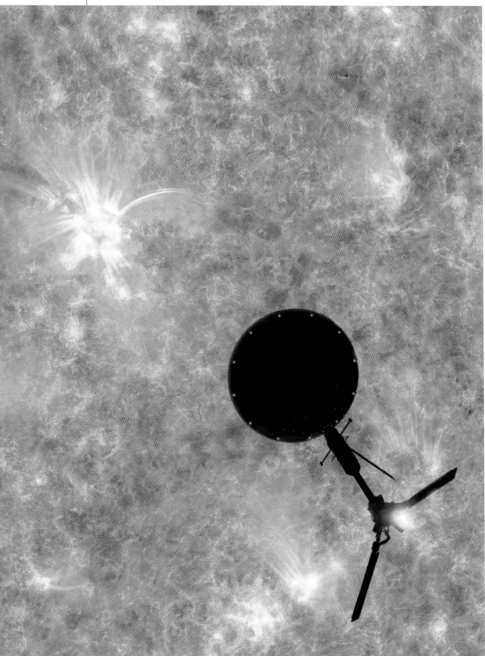

Pegasus **at its closest approach to the Sun.**

Privately, I've doubted the wisdom of this trajectory for some weeks now, ever since the solar flare that nearly wrecked our Mars mission. If the same thing happened now, we'd be toast. I was present at the planning meetings that decided our course, and I remember thinking that some of the scientists, who would watch us fry from afar, had been a little too keen on the Sun burn. I know the route makes time-saving and fuel-saving sense; and it will deliver on science. But there's no fail-safe on this manoeuvre, no scientific or technological protection – we are just riding our luck, flying on a wing and a prayer.

myself imagining that this great seething mass was somehow aware of our presence and might suddenly stir and lash out at us.

Mars, the Moon, and even Earth have faces that hardly change as you watch them. Their landforms are static and predictable; only the imperceptible drift of clouds can slowly change the view. But here above the Sun our view was of a distorted and twisted face of constantly shifting features. It was visibly alive – and on a grand scale. Cracks hundreds of thousands of kilometres long would open up and slam shut within hours. Gnarled streamers of plasma arched from the surface in magnetically tethered loops that snapped into spiky forests of flailing arms. Tugged this way and that by powerful magnetic force fields, the arms would then rejoin again moments later. Eruptions vomited out glowing gases that gathered into Earth-sized twisters and danced demonically across the surface. It all felt to me like some disturbing scene of pain and anguish, as cruel forces inside the Sun tried to stab and slash their way out. Injuries would heal, only for the new skin to be torn open and tormented once more as the surface writhed in an endless cycle of suffering and misery.

High above this restless surface, *Pegasus* was under attack. A blizzard of gamma-ray photons, high-energy X-rays, and super-accelerated protons and electrons bombarded the outside of the ship, sending sparks dancing along the truss and lighting up the poles of our magnetic shield. Camera images constantly crackled with interference, and radiation sensors outside skipped into their upper scales. Minor computer systems, tripped by rogue

particles, froze and defaulted to safe mode. Yvan cursed his way through each re-boot, grumbling about the computers' poor performance. It was hard not to get paranoid about this invisible threat seeping into our shelter. The hairs on my arms began to stand up, and I started to smell ozone in the command centre as the oxygen around us was ionized. I kept glancing at my dosimeter, expecting to see it glowing amber or red.

Hurtling through the Sun's corona, cocooned in our tiny capsule, was a claustrophobic experience. My chest tightened and my heart pounded with anxiety. The audacity of sailing this close to a star felt way beyond recklessness and perhaps even entered the realms of insanity. This wasn't a place for life. It was a death zone, and we were getting closer and closer.

Right on target, and four milliseconds early, we soared past our closest point to the Sun and hit a top velocity of 315 km per second – 1.13 million km/h! The Sun immediately began to contract on the monitors as the gravitational slingshot propelled us away from it, and a sudden low rumble signalled that our main engine had fired right on cue. I heard everyone let out a sigh of relief.

We dropped several 'suicide probes' during the flyby, and these will help scientists map the corona in three dimensions and plot magnetic fields and plasma-wave intensities. The probes have already recorded solar wind speeds of 400 km per second at the equator. And just after perihelion, all our cameras picked up the ripples of a 'sunquake'. We don't know what caused it, but – fortunately for us – the epicentre was on the far side of the Sun.

———————

✪ Tom Kirby, Commander
204 days, 11 hours, 23 minutes

About 16 hours after perihelion, radio blackout ended and we re-established contact with Earth. One of the first messages to come through was from Space Environment Centre – apparently a small solar flare emerged on the western limb of the Sun during our closest approach. That's what triggered the sunquake we detected on our side. I guess I'm glad we didn't find out about this until afterwards.

The gamble seems to have paid off, and we are now on a course for the outer solar system. Extraordinary though the solar flyby was, I am so, so glad to be heading away from the Sun and onwards to Jupiter.

———————

Tension at Mission Control during the solar flyby.

Satellite Sun-seekers

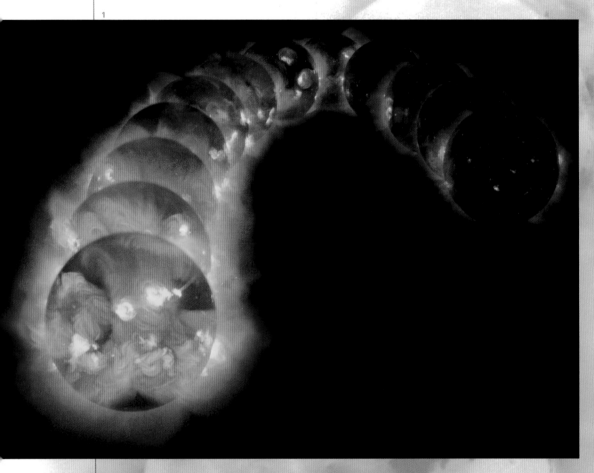

and Earth's magnetic field. In 1962 NASA launched the first of a series of Orbiting Solar Observatories to watch the Sun in ultraviolet, X-ray and gamma-ray wavelengths for an entire 11-year solar cycle.

Ten years later, both superpowers had manned missions 'to' the Sun. NASA astronauts were observing the Sun from the *Skylab* space station in orbit around Earth, and Soviet cosmonauts were doing the same from *Soyuz 4*. The Soviets also launched a series of Prognoz satellites to monitor the solar wind, and as luck would have it they registered a huge blast of solar wind within a few days of launch in 1972; it may have been the first recorded coronal mass ejection. Further satellites studied the solar wind in the 1970s, and in 1980 NASA launched the *Solar Maximum Mission* spacecraft to coincide with the peak of the solar cycle.

By the turn of the century there was a new generation of spacecraft in place to watch the Sun. The joint European Space Agency/NASA probe *Ulysses* was launched in 1990 from the space shuttle *Discovery*

Today there are more than 20 solar-physics missions in flight or about to launch, which means the Sun is currently the most studied object in the solar system. From the comfort of your armchair, safe inside Earth's magnetosphere, you can log on to the internet and watch solar storms rage. You can check today's solar weather forecast, and you can find out how much radiation will be thrown at our planet in the next few hours. But just 50 years ago, blinkered by our magnetic bubble, we were completely blind to all this.

One of the first spacecraft to study the Sun was the USSR's *Lunik 3*. In 1959, while

Lunik 3 was on its way to take the historic first picture of the far side of the Moon, its sensors detected an unexpected surge in the level of charged particles flying through space. It was the first concrete evidence of the solar wind – the mixture of electromagnetic waves and charged particles that flows out of the Sun all the time. The solar wind can vary enormously in strength, depending on the Sun's activity.

More spacecraft followed in the 1960s, including NASA's Explorer and Pioneer probes and the USSR's Electron satellites, all of which carried instruments to study the interaction between the Sun's radiation

and used Jupiter's gravity to slingshot itself out of the plane of the solar system and into a solar polar orbit, providing our first observations of the Sun from above and below. After initial teething troubles that caused the craft to wobble critically, *Ulysses* reached Jupiter in February 1992 and has since made four giant polar orbits of the Sun at a distance of about 1 billion

board can detect the faint tremors of sunquakes, which are triggered by solar flares.

After its first launch went disastrously wrong in 1996, *Cluster* – a constellation of four identical spacecraft that fly in formation to map the solar wind in three dimensions – was rebuilt and relaunched in 2000. *Cluster* orbits Earth, passing in and

1. The solar cycle spanning 11 years; from solar maximum (on the left) to solar minimum (on the right) taken by the *Yohkoh* satellite.
2. Model of the Soviet *Lunik 3* spacecraft – the first attempt to study the Sun from above our atmosphere.
3. Gnarled magnetic field lines break through the Sun's surface.
4. A *SoHO* image of the Sun taken using the Extreme Ultraviolet Imaging Telescope. It's a million miles across.
5. A minor solar eruption caught by the spacecraft *TRACE*.

km (620 million miles), providing the first map of the poles.

Another joint NASA/ESA craft – *SoHO* (*Solar Heliospheric Observatory*) – was launched in 1995 and has been perhaps the most successful solar observatory to date. *SoHO* lies directly between Earth and the Sun, at the point where the Sun's gravity and Earth's are matched. This keeps *SoHO*'s orbit of the Sun locked in step with ours but also gives it an uninterrupted view of the Sun – something that would be impossible for an Earth-orbiting satellite. *SoHO* returns daily images of the Sun in a variety of wavelengths, providing an advance warning of any solar activity that might threaten Earth. One instrument on

out of our magnetic field and investigating how the Sun and Earth interact.

Among the armada of other spacecraft watching the Sun are *TRACE* (*Transition Region and Coronal Explorer*) and *Yohkoh*, both of which provide spectacular time-lapse movies of the seething solar surface. Even NASA's Voyager probes, which were launched in 1977 and are currently in the outer reaches of the solar system, will add to our knowledge of the Sun. The *Voyagers* are the farthest man-made objects from Earth, and in a few years they will cross the 'heliopause' – the outer edge of the bubble created by the solar wind. Many scientists consider this to be the outer boundary of the solar system.

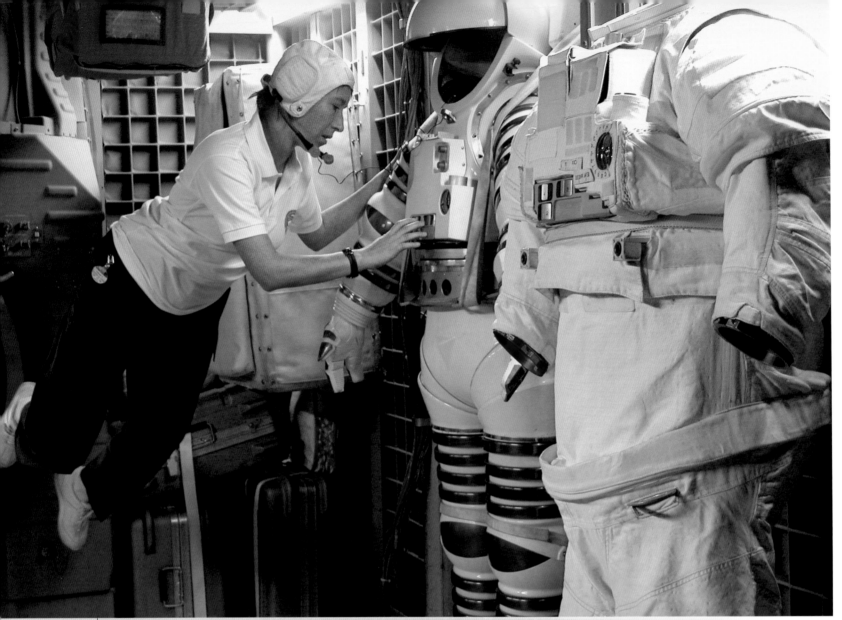

Lessard floats in the *Pegasus* airlock, checking her Io-EVA suit.

➕ **John Pearson, Flight Medic**
204 days, 9 hours, 12 minutes

I know *Pegasus* is well protected against radiation, and the magnetoshield apparently performed flawlessly, but we just passed within 8 million km (5 million miles) of a billion-trillion-trillion-tonne nuclear engine. Radiation levels outside were more than a hundred million times the lethal level for a human. Life and stars just aren't compatible, and no matter how much shielding you've got, it's my belief that astronauts should stay away from them. Tom reported smelling ozone in the command centre, but I think his imagination was playing tricks on him. It would have meant a serious leak, but none of the bubble tubes registered such a breach. From our personal dosimeters

I'd estimate that we took the equivalent of about 40 chest X-rays during the flyby, which is about five times the normal daily dose up here. It's about what we'd predicted. With the extra antioxidant supplements in our diet, there should be no long-lasting effects.

I confess that my own dosimeter seems to have edged into the amber. I'm running some tests to see if this is a real reading or some kind of error. For most of the flyby I was sitting in the seat to the right of Tom, and there's nothing else to suggest a leak got into this part of the command centre (unless Tom was right about the ozone smell). If the dosimeter is correct, I probably took a dose equivalent to 200 chest X-rays. I'll be upping my antioxidant supplements as a precaution.

Less well protected parts of the habitation module suffered more than the command centre. The computers in the lab area all failed, and some of the plants growing in there seem to be suffering from brown marks on their leaves. Yvan tells me there are urgent repairs to do on the outside of *Pegasus*, and he's already planning the EVAs. I'm sometimes glad I've only got four humans to worry about.

☒ **Yvan Grigorev, Flight Engineer**
230 days, 12 hours, 30 minutes

Since the solar flyby, everyone has been busy repairing the radiation damage to equipment in the habitation modules, running checks on recycling and life-support systems, and carrying out daily maintenance programmes. We are now 28 days and just under 160 million km from the Sun – about 1 astronomical unit, which is how far Earth orbits. We're climbing out of the Sun's gravity well, and the forces that made us speed up during our fall have now slowed us to less than 200,000 km/h. But we would be travelling a lot more slowly if we hadn't made that engine burn at perihelion.

Now that the Sun is far behind us, it's safe to work outside *Pegasus*. I've already conducted two of four planned EVAs to make observations and repairs to the spacecraft superstructure. One of my principal tasks was to inspect the main shield. It's made of a heat-resistant material called reinforced beryllium-alloy, which is covered with carbon foam and then finished with a 20 millimetre layer of beryllium. During the solar flyby, heat and radiation stripped off parts of the beryllium layer. The shield appears to be working, but I need to carry out a more careful inspection of its weakest areas, where equipment bays and structural points are located. In another month this shield will be turned forward again to protect us from micrometeorite impacts as we cross the asteroid belt, and five months later it will scrape the top of Jupiter's atmosphere to slow us down as we enter orbit. Our lives depend on its strength and integrity.

Working deep in the shadow of the shield, I sometimes turn off my lights and pause to look up at the stars. There are more than you can ever see from anywhere on Earth. Alone out there, dwarfed by the vast *Pegasus* superstructure, I sometimes feel the ship is alive. I can feel the vibrations through the handrails as the long truss flexes with the rotation of the sleep compartments almost a kilometre away. Behind me a constant tremor emanates from the reactor, amplified by the cavernous space inside the shield disc. My ship is a true leviathan. It feels alive in a way that only giant mechanical constructions can.

👁 **Zoë Lessard, Mission Scientist**
240 days (5 days to asteroid belt entry)

Jupiter draws nearer, and with it the moon Io, where I will make my first EVA. It's what drives me on through those hours of tedious exercising.

But before *Pegasus* can reach Jupiter, it must cross the asteroid belt, a band of lonely rocks floating in space between Mars and Jupiter. Only a geologist could love them. Some think they are leftover debris from the formation of the solar system, prevented from forming a planet by the gravitational tug of Jupiter. Astronomers reckon the belt contains about 1.4 million asteroids, less than 4 per cent of which are charted. Some 16,000 are named, and we know they pose no threat to the flight.

The belt is spread out across such a vast area that the fragments, which range from dust particles to rocks the size of France, are rarely in eye-shot of each other and should therefore present little danger to a transiting spacecraft. That said, *Pegasus* has a radar warning system that scans some half a million kilometres ahead of the craft and can give us a 2.5-hour warning of an impending impact. I guess it's a little like crossing a minefield using a very long stick to check the ground ahead.

In the unlikely event that we find ourselves on a collision course with an asteroid, it should take only 20 minutes to collect enough data to calculate an emergency mid-course correction to avoid an impact. However, protocol requires that we confirm any such manoeuvre with Control, and that could take as long as 45 minutes. This cuts it fine, but Control always reassure us that they can react fast.

Things have been very quiet for the last month, but once we enter the asteroid belt we will have the added excitement that we might be heading for a brick wall the size of New York. Who said interplanetary space travel was dull?

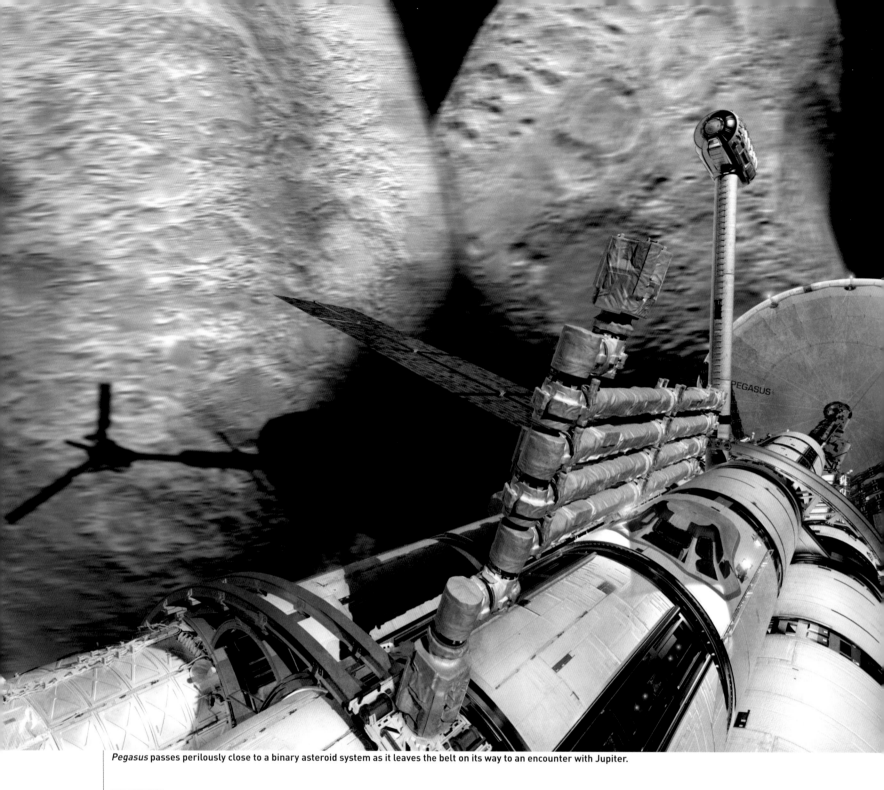

Pegasus passes perilously close to a binary asteroid system as it leaves the belt on its way to an encounter with Jupiter.

☒ **Yvan Grigorev, Flight Engineer**
273 days, 17 hours, 26 minutes

Despite Control's endless assurances that we are extremely unlikely to even see an asteroid, let alone hit one, there's been an unspoken tension since we entered the asteroid belt four weeks ago. I think we are all mentally counting down the days until

Pegasus emerges from the other side. Dust impact on the main shield, now at the front of the ship, has risen marginally during the crossing, but it's the large fragments we're worried about. At the speed we're travelling, anything larger than a tennis ball could do serious damage to the shield.

Although we all trust the early warning system, it's difficult not to feel uneasy when we go to sleep at night. And there's always the possibility of being jerked awake by the alarm.

⊠ Yvan Grigorev, Flight Engineer
280 days, 17 hours, 40 minutes

Against all the odds, we had a very close encounter early this afternoon. Nina, Zoë and I were in the food store unpacking reserves for the week ahead when the alarm went off. We scrambled back to the command centre hoping Tom was running a drill.

According to the warning system, there was a 99 per cent chance of a collision with a sizeable object 126 minutes ahead. I quickly ran the flight correction options through the computer and calculated that a 30-second burn from all eight positional thrusters on the underside of *Pegasus* would be enough to avoid impact by a wide margin. If we acted quickly, the asteroid would miss us by 300 km. But the longer we left our decision, the longer we'd have to fire the thrusters to achieve this miss distance.

Tom insisted on me re-checking my figures before he requested permission for the burn from Control. It took another minute to run the simulations again and I came to the same conclusion. Tom sent the parameters back to Earth. Control had already been alerted by an automatic message from the warning system, and they were working out their own solution. It annoyed me that we had to wait for their blessing.

The countdown to impact had reached 81 minutes by the time the answer came back. I was stunned: they were advising us to make no burn. Apparently, extra measurements had been taken into account, and they reckoned we'd miss the asteroid by just over 10 km without having to make a burn. A close flyby would also allow us to snatch some extra science from the mission. Control reckoned the asteroid was a C-type, and there was something strange about the radar data that made them keen for a closer look.

I ran the simulation again with Control's new parameters. I felt it was too close for comfort, and I stood by my recommendation to make the burn. But Tom was reluctant to discuss it, preferring to stand by Control's decision. I thought he was being boneheaded, but I suppose he trusted their figures and felt it would be wrong to overrule them. There was silence in the cupola as we crowded in to watch the asteroid sweep past at what we trusted would be 10 km away.

✳ Tom Kirby, Commander
280 days, 18 hours, 45 minutes

Our unplanned flyby of the two asteroids, provisionally named 30241 Hubris and 30242 Catastrophe, occurred 13 hours and 6 minutes into day 280 of the mission. We passed within 950 m of the surface and recorded the flyby on all available cameras and spectrometers. As we drew close, radar reflections suggested there might be two asteroids, and closer observations confirmed this. The primary body was 16 km long and the smaller one was 8 km and possibly U-shaped. Further measurements of the smaller body couldn't be made as the larger one blocked our view. The pair were locked in orbit around each other, and we estimated their centres to be 3.4 km apart and their orbital period to be 94 hours.

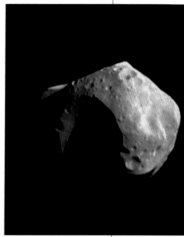

Spectral observations suggest that both were indeed C-type asteroids, with dark, reddish surfaces. Such bodies are made of the same 'carbonaceous chondrite' rock that we think makes up most of the rocky bodies in the outer solar system. However, there were perturbations in the magnetic field as we passed the asteroids, which suggests they contain some kind of magnetic material not typical of C-types. Their gravitational influence on our own trajectory implied a density of only 2.8 g/cm³, which suggests their insides are probably severely fractured and honeycombed. The visible surfaces of both were heavily cratered – pulverized even.

Though I am sure the encounter is being portrayed on Earth as another triumph for *Pegasus*, it was a disaster up here, especially for morale. Until Hubris and Catastrophe, our trust in Control was absolute. But now there is doubt. Yvan was right all along, and all the hundreds of people at Control, with all their computing power, were wrong. Maybe it was just a bunch of backroom science guys getting greedy, or worse still, maybe it was incompetence. Our survival on this mission is in the hands of flight controllers on Earth, and they have no right to take unnecessary risks. It's shaken my faith in Control's advice, and with so far still to go on this flight, that's not a good thing. I won't be making the same mistake again.

A close encounter with an asteroid.

Through the Sun's Corona

89

The Brightest Sky

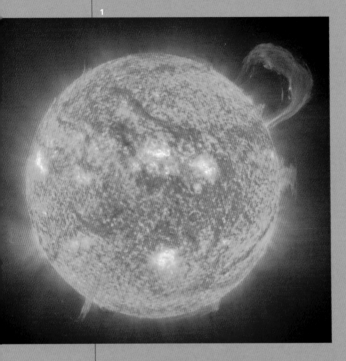

Outside the Sun's visible surface is the chromosphere – an irregular, darker layer that is visible only during a solar eclipse. At around 20,000 °C (36,000 °F), the rarefied hydrogen in the chromosphere is actually hotter than the Sun's surface and has a reddish glow; hence the layer's name, which means 'colour sphere'. Tongues of hydrogen often detach from the photosphere and rise into the chromosphere to form glowing loops of gas, or solar prominences.

The chromosphere is where solar flares erupt from. These explosive bursts of gas happen when arching magnetic field lines collide in the chromosphere. Small flares are a daily occurrence, but larger ones – called coronal mass ejections or sun storms – happen only occasionally. CMEs can be catastrophic. In just a few hours they release as much energy as many billions of

megatons of TNT. A storm of high-energy particles and radiation is hurled into space – sometimes in our direction. When the particles hit Earth they brighten the auroras and can damage satellites or even ground stations.

Beyond the chromosphere is the Sun's outer atmosphere, or corona, which is visible as a white halo around the Moon during a total eclipse. The corona expands and contracts during the solar cycle, and – bizarrely – it is even hotter than the chromosphere, with temperatures of 1–2 million °C (2–4 million °F). Quite how this region could be hotter than the layer below has proved hard to explain. Physicists suspect that energy is somehow transferred along the magnetic field lines that loop into the corona. According to some estimates, each loop carries about as

much energy as the Hoover Dam can generate in a million years.

Beyond the corona is the heliosphere – a gigantic bubble created by the solar wind, which streams out of the Sun and across the solar system at 350–750 km (220–470 miles) per second. The heliosphere defines the full extent of the Sun's influence and reaches way beyond the orbits of the planets, maybe 40–50 times farther from the Sun than Earth. Like the corona, it expands and contracts with the Sun's cycles.

1. An extreme ultraviolet *SoHO* image of the Sun captures an immense prominence or solar flare arching 250,000 miles above the surface in 1999.
2. The 2-million-degree solar corona captured during a total eclipse of the Sun in July 1991. The image was taken from Baja, California.

The Missing Planet

There's a gap in the solar system between Mars and Jupiter. To astronomers of two centuries ago, this gap was a mathematical problem in the pattern of spacing of the planets. So, in 1800, the German astronomer Johann Schroeter assembled a team of six 'celestial police' to track down the missing planet.

The mystery seemed to be solved when Giuseppe Piazzi, director of the Sicilian Observatory, found a planet on 1 January 1801. But Ceres, as it was christened, was only 933 km (580 miles) in diameter, so the celestial police continued their search. Over the next six years they found three more tiny planets – Pallas, Juno and Vesta – but all were even smaller than Ceres. Even so, the gap in the solar system was filling up.

What the astronomers had stumbled across was the asteroid belt – a zone of rocky debris that orbits the Sun between the orbits of Mars and Jupiter. There are now 16,000 named asteroids, and more than 48,000 have been catalogued. Ceres is the largest, accounting for a quarter of the mass of the whole belt. We've probably found 99 per cent of asteroids larger than 100 km (62 miles) in diameter, and about half in the 10–100 km (6–62 mile) range. But there may be more than a million smaller asteroids waiting to be discovered.

Quite why all this debris didn't form another planet is still something of a mystery, but Jupiter's gravitational influence on this part of the solar system is thought to be the main culprit – preventing another planet from growing. It's likely however that some of the asteroids we see today were once part of larger rocky worlds out here – which were later broken apart by collisions. The resulting chunks

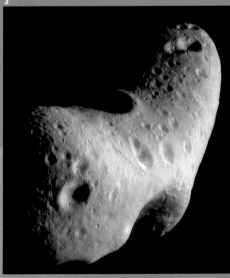

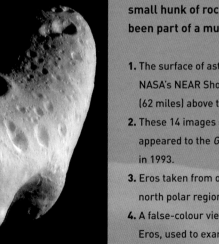

varied in their chemistry and geology enough to form the basis for a simple classification system used today.

The *Pioneer 10* and *11* and Voyager probes were the first spacecraft to cross the asteroid belt, and much to the relief of mission control, they came nowhere near any asteroids. Later missions across the belt made deliberate close encounters. In the early 1990s *Galileo* flew past two asteroids: a C-type called 243 Ida and an S-type called 951 Gaspra. Gaspra was found to be a mixture of rock and metal, with a heavily cratered surface suggesting an age of maybe 200 million years. Ida even had its own tiny moon – a 1.6-km (1-mile) long asteroid named Dactyl.

NASA's *Near Earth Asteroid Rendezvous* (*NEAR*) probe was even more successful than *Galileo*. In 1999 *NEAR* went into orbit around 433 Eros, and in 2001 it landed on the asteroid's dusty surface. The beauty of its cratered landscape, covered in grooves and house-sized boulders, surpassed anything that mission controllers had dreamt of seeing on such a small hunk of rock. It seemed to have once been part of a much bigger world.

1. The surface of asteroid Eros captured by NASA's NEAR Shoemaker mission, just 100 km (62 miles) above the ground.
2. These 14 images show an Ida day as it appeared to the *Galileo* spacecraft's camera in 1993.
3. Eros taken from orbit looking down on the north polar region.
4. A false-colour view of the surface of asteroid Eros, used to examine the amounts of space weathering on the surface.

Here Be Giants

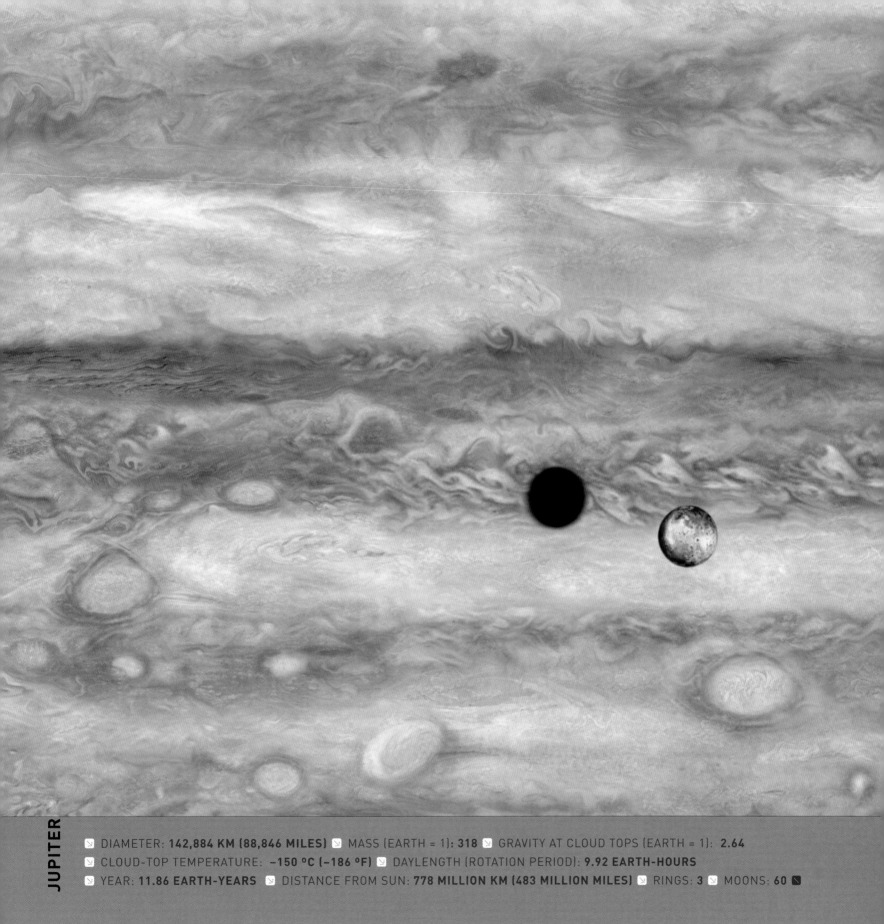

JUPITER

⬚ DIAMETER: **142,884 KM (88,846 MILES)** ⬚ MASS (EARTH = 1): **318** ⬚ GRAVITY AT CLOUD TOPS (EARTH = 1): **2.64**

⬚ CLOUD-TOP TEMPERATURE: **−150 ºC (−186 ºF)** ⬚ DAYLENGTH (ROTATION PERIOD): **9.92 EARTH-HOURS**

⬚ YEAR: **11.86 EARTH-YEARS** ⬚ DISTANCE FROM SUN: **778 MILLION KM (483 MILLION MILES)** ⬚ RINGS: **3** ⬚ MOONS: **60** ◩

The mission had come very close to disaster when the navigation miscalculation nearly sent *Pegasus* slamming into asteroid 30241 Hubris. Privately, Commander Tom Kirby blamed the science team for being greedy and trying to squeeze ever more data out of the mission, but the truth is that the decision not to make a course correction came from flight guidance, not science. And those were five of my closest friends up there – I would never have done anything to endanger them. However, looking back, I do feel some guilt that our enthusiasm might have influenced the guys in flight guidance. They were heady days – everything had gone so well so far, and we had felt invulnerable. But the near miss was a wake-up call.

After Hubris, the crew's faith in us was shaken. It was bad enough that they thought we'd made a mistake, but now they started to think we saw them as expendable. Trust was essential if this mission was to succeed, and we'd damaged it before getting even a fifth of the way through. And once lost, trust is difficult to regain – all of us at Control had a tougher task from that point on.

On the plus side, Jupiter was drawing near and started to consume everyone's attention. I first read of a manned Jupiter mission in Arthur C. Clarke's novel *2010* when I was a boy; little did I know I'd be involved in a genuine human mission to this gigantic ball of gas some 40 years later. Like much else that Clarke foresaw in his fiction, our ship also used Jupiter's enormous atmosphere to slow its approach, in the most daring aerobraking manoeuvre yet attempted.

From the moment *Pegasus* entered Jupiter's magnetic field, or 'magnetosphere', the clock began ticking. The crew had to drop a probe into Jupiter's atmosphere, land on the moon Io, and collect samples from the frozen oceans of Europa – a moon harbouring twice as much water as Earth's oceans. All this had to be done in less than a month to minimize exposure to Jupiter's radiation belts. It was a close call, and the deadly radiation almost finished the mission.

Alex Lloyd
Chief Scientist, *Pegasus* Mission

✚ John Pearson, Flight Medic
Mission elapsed time: 370 days, 12 hours, 9 minutes

We've been acquiring X-ray and radio emissions from Jupiter for the past two days to get a picture of the planet's magnetosphere. It's enormous, and to be honest, I think everyone is a bit intimidated. It's big enough to engulf all Jupiter's 60 moons, and its tail, blown out into space by the solar wind, could touch the clouds of Saturn 800 million km away. We are painfully aware of the radiation trapped within the magnetosphere, and this makes it a very threatening object.

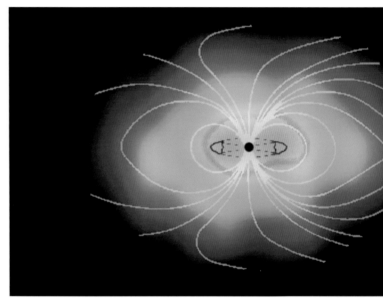

Jupiter's magnetosphere – the largest object in the solar system. Almost 2 million km across (and with a tail up to 650 million km long).

Four days before we hit the cloud tops, we'll cross the 'bow shock' – a wall of pressure where the force of the solar wind meets the force of the magnetosphere. From there in we'll encounter heightened levels of radiation, as charged particles snatched from the solar wind are propelled to lethal speeds. The dangers will climax three hours before we crash into the atmosphere, when we pass through Jupiter's vicious Van Allen belts. Like Earth's belts of the same name, these are where magnetic field strengths reach a peak.

The radiation makes a stay near Jupiter almost as dangerous as our solar flyby. But unlike our brief brush with the Sun, our stay in the Jupiter system will last weeks – and the longer we linger, the higher the risk of a lethal dose. Jupiter's volcanic moon Io lies within the Van Allen belts, and it'll be a race against time to explore its exotic surface. I'm introducing a new radiation medication as of tomorrow and boosting dietary supplements as well.

All these magnetic fields, waves of radiation and streams of gyrating electrons pump out a cacophony of radio signals. Yvan has been picking them up and playing them through the speakers around the hab module. Perhaps it's because I'm the doctor, but I can't help thinking the howl of these deadly rays is nature's warning to stay away from Jupiter. Ironically, it's also what makes Jupiter so interesting and compels us to visit – a sort of scientific Siren call.

If the magnetosphere wasn't enough to put up with, we've also got a hazardous aerobraking manoeuvre to pull off if we're to

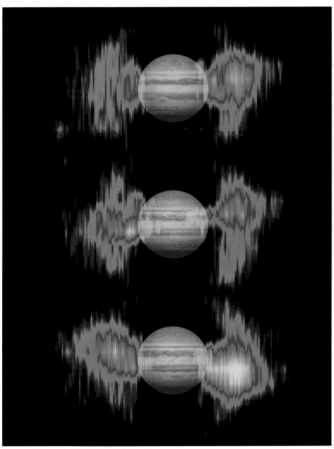

The radiation belts generated by the magnetic field wobble as the planet rotates.

enter orbit. Crashing into Jupiter's cloud tops will make us decelerate so powerfully that we'll weigh up to nine times our normal weight. The G-suits and the couches in the command centre will help us withstand the deceleration, but we need to be at the peak of physical fitness to come through unscathed. I've boosted our daily exercise programme to five hours a day. Our bodies are in for one hell of a beating.

🔋 Nina Sulman, Mission Scientist
385 days, 7 hours, 20 minutes

Fifteen days from Jupiter and we're well and truly locked into its tractor beam, its gravity hauling us closer. For the first time in months, our speed began to pick up this morning as Jupiter's gravity finally began to outweigh the Sun's. Jupiter's gravitational influence is colossal. It has pulled in passing comets and asteroids for billions of years, ensuring that many on a deadly collision course with the inner solar system never make it to their destination. In fact, some believe that life on Earth would have taken a very different course if Jupiter wasn't hoovering up so much of the deadly debris that hurtles towards us.

The moons Callisto (left) and Europa, closer to Jupiter.

And now Jupiter is pulling us in too. Already, the familiar stripes of this 'gas giant' are clearer than I've ever seen them. It's sometimes said that if Jupiter had been a little bigger it would have turned into a star. Well, it's no match for the Sun, but it's the biggest planet in the solar system by far, with enough mass to have trapped its own suite of orbiting worlds. With at least 60 moons caught in Jupiter's gravity trap, the Jovian system is like a solar system in miniature. Perhaps we'll discover a few more moons on our visit. Most of the smaller ones are difficult to see – tiny captured asteroids that strayed too close. But clearly visible for several weeks now are the four main moons, some of which are bigger than planets. These are the Galilean moons, first discovered by the Italian astronomer whose name they bear. We are carrying a sketch he made of them in 1610 as he charted their dance around Jupiter, proving that not everything orbited the Earth. It was a gift from the Vatican. They asked Zoë to carry out a little ceremony and lay it on Io; she seems quite happy to, but to me it seems a bit odd.

We've all done a lot of training for Zoë's landing on Io. There's plenty to find out about this amazing but puzzling moon. Ever since *Voyager 1* discovered volcanic plumes the size of Texas

Io's volcanoes are interesting for another reason. They produce a strange kind of lava that was only around on Earth very early in our planet's history, when Earth was hotter and more geologically active. It's a rare opportunity to study primeval volcanoes as if we were travelling back in time on Earth. It's a geologist's dream, and I can tell Zoë is excited.

But such an opportunity comes at a price. Io orbits Jupiter inside a deadly cloud of highly energized charged particles. Radiation levels on the surface can rise to 500 times the lethal dose for a human, and without protection Zoë would be dead in seconds. Io probably isn't the best place to look for life, but there's somewhere else in the Jovian system that might be more hospitable: Europa, Jupiter's second Galilean moon.

Europa is covered in water ice, and some experts think it's a better place to hunt for life than Mars, and on the basis of our luck at Mars, I think I'd agree with them. Close inspection of Europa by the *Galileo* spacecraft in the 1990s revealed intriguing

Jupiter's rainbow moon Io – the colours are due to sulphur compounds.

An eruption from the volcano Prometheus catches the sunlight.

erupting from the surface, geologists have argued over how such a tiny world could be so volcanically active. It should have frozen to death billions of years ago, but its interior is hot enough to drive tremendous daily eruptions that alter the landscape as quickly as clouds alter the sky back on Earth. Part of the heating must come from the gravitational tug of nearby Jupiter, which stokes fires deep inside Io by causing friction. But some eruptions throw out more energy than could be accounted for by tidal heating alone. Something else must be going on, and I hope we can find out what.

patches of colour in the ice. Some think the colour might come from microbes brought to the surface by upwellings of warm water rising from deep volcanic vents. To get any useful samples from Europa, we'll need to drill through the surface. This would have taken us too long, so an advance probe has been down there for some months now – boring through the floor of a crater to retrieve deeper samples. *Pegasus* will then rendezvous with it to retrieve the samples. I can't wait to get my hands on them.

💬 Zoë Lessard, Mission Scientist
400 days, 9 hours, 30 minutes

Tom and John keep asking if I'm worried. I wish they'd shut up. Of course I'm worried! I want to do well, and I'm determined this mission won't fail because of me. I'm training six hours a day to keep myself strong and fit. The lander is built with the toughest radiation-proof materials, and so is my spacesuit. I'll even be carrying my own personal magnetic shield. And as if that wasn't enough, I'll be landing on the side of Io that isn't taking the full brunt of the Van Allen belts. I'll be just fine.

Every spare moment I'm not exercising or sleeping, I'm running through the exploration plans for each of the possible landing sites. I've even got a flight plan in my hand at meal times these days. Control decided not to choose the final site until we're in orbit, so I've got to keep up to speed on all of them. Almost all Io's volcanoes seem to be warm, so it's hard to say which are dormant, and we won't be sure of the stability of a particular site until we're much closer. So for now, there are four possible scenarios, and all but one of them puts me close to an exotic Ionian volcano.

I've been watching my little moon for the last couple of months with my own eyes. We are the first people to see it this way – unaided, unmagnified. The ochre surface grows clearer each day. I can already make out the biggest mountains and volcanoes. It must be the most variegated surface in the solar system – sulphurous hues of so many shades, from dirty greens through brick reds and browns to oranges and the palest of yellows. Volcanic plumes arch over the horizon, glowing with a blue aura. There must be a hundred tonnes of material bursting out of these eruptions every second. That's way more than any volcano on Earth. Through binoculars I can make out active lava flows pouring from lakes so hot they glow white. At these temperatures the lava gushes with the fluidity of cooking oil. Elsewhere there are frozen lakes of solidified lava, their glassy surfaces glinting in the weak sunlight. None of the danger is visible from this distance, just the utter beauty of this unique and precious moon.

Jupiter's cloud patterns west of the Great Red Spot.

❌ Yvan Grigorev, Flight Engineer
406 days, 16 hours, 3 minutes

Jupiter's massive, striped face dominates all the screens in the command centre. We are now five days away from the dreaded aerobraking, and after just four course corrections en route to Jupiter, we are within 800 m of our planned position.

But the tricky bit comes when we hit Jupiter's upper atmosphere to slow *Pegasus* down. We have to enter the atmosphere at exactly the right angle, travelling at more than 30 km per second. The margin for error is tiny. A fraction too steep and we risk burning up; a fraction too shallow we'll skip off like a stone bouncing on a pond – and off into deep space on an uncorrectable course. There can be no missed zeros or minuses, no mix-ups of imperial and metric. Until a few weeks ago I wouldn't have worried about Control's ability to place us on the right trajectory, but after Hubris, who knows. I find myself checking and rechecking their calculations – it's very draining.

Jupiter isn't a place to take a craft that's already a veteran of missions to Venus, Mars, and the solar corona – not to mention a close shave with a binary asteroid. Failure of the computers that control the timing of the events during the aerobrake could be disastrous.

King of the Heavens

Jupiter was one of the five 'wandering stars' known to the ancients. It grows very bright during the year and is outshone only by Venus. But unlike Venus, which is close to the Sun and visible only around dusk and dawn, brilliant Jupiter rules the late-night skies. The Ancient Greeks and Romans named it after their most powerful Gods: Zeus and Jupiter.

The first telescopes could make out a banded disc accompanied by four pinprick dots of light – the Galilean moons. These were discovered by the Italian astronomer Galileo, who sketched their movements around Jupiter in 1610. A year later, the German astronomer Simon Marius claimed to have seen them almost two months before Galileo, and a heated debate ensued, with Galileo accusing Marius of copying his work and backdating it. Marius lost his place in history, but it was his idea to name the moons after Jupiter's lovers in Greco-Roman mythology: Io, Europa, Ganymede and Callisto. Incredibly, the names were not officially accepted until three and a half centuries later, by which time two spacecraft had already flown past Jupiter.

Meticulous charting of the moons' orbits revealed a very predictable pattern, but when Jupiter was at its farthest from Earth, the moons seemed to run slow. In 1675 the Danish astronomer Ole Roemer suggested the delay was caused by the finite speed of light, which he calculated as 225,000 km (140,000 miles) per second – a good estimate, but slightly short of the true figure of 300,000 km (186,000 miles) per second.

Improvements in optics made it possible to measure Jupiter's diameter and study the bands of colour that span the planet, though it still wasn't known whether these represented a solid surface or just weather patterns. In 1664 Robert Hooke saw a large oval spot. Other spots were soon noted and used to calculate the speed of Jupiter's rotation, but astronomers had little idea what the spots were. Most of them came and went, deepening the mystery.

In 1687 Isaac Newton used his theory of gravity to work out the masses of the planets and moons. Given the speed that Jupiter orbits the Sun, and the speed that the Galilean moons orbit Jupiter, Newton reckoned Jupiter weighed only 318 times Earth's weight, despite being some 1300 times greater in volume. The planet had to be made of gas, which meant the markings on the surface were weather systems – maybe even storms.

Telescopes improved in the nineteenth century, and astronomers discovered that Jupiter's markings move around the planet at different rates. It was becoming clear we were watching the top of a very turbulent atmosphere. But if the spots were storms, they were unlike any storms on Earth as they seemed to rage for years at a time. The Great Red Spot may have existed for centuries, though it apparently disappeared in the late eighteenth century and was not seen again until 1840. We're still unsure if it really did disappear, but when it 'returned' it had changed in colour and size.

The first clues to the planet's composition came in 1903, when US astronomer Vesto Slipher turned the new science of spectroscopy from the stars to Jupiter. But the spectral lines that show which elements are present looked strangely blurred for Jupiter. It took a fresh pair of eyes to interpret these smudges as compounds rather than pure elements, 30 years later. First ammonia and methane were detected, and later hydrogen.

In the 1950s the German–American astronomer Rupert Wildt came up with a model of Jupiter's interior that is still generally accepted. Unlike the rocky inner planets, Jupiter is a 'gas giant' made up largely of hydrogen and helium left over from the Sun's formation. The swirling clouds are not part of a thin atmosphere around the planet but the top of an ocean of gas many thousands of miles deep. Near the centre of Jupiter, gravity compresses the hydrogen into a dense, liquid form that acts like metal and can conduct electricity. And right in the centre of this there is probably a rocky core perhaps 15 times the mass of the Earth – the original seed-planet around which Jupiter formed.

Around the same time as Wildt's work, radio astronomers Bernard Burke and Kenneth Franklin stumbled across a very powerful radio signal coming from Jupiter. Wildt's churning core of metallic hydrogen was generating a giant magnetic field that trapped charged particles from the solar wind, causing them to emit radio waves. If we could see it from Earth, this giant magnetic bubble would appear up to five times as big as the full Moon.

1. Jupiter and the four Galilean moons seen from Earth through a telescope.
2. Galileo's drawings of orbits of the moons of Jupiter made in 1613.

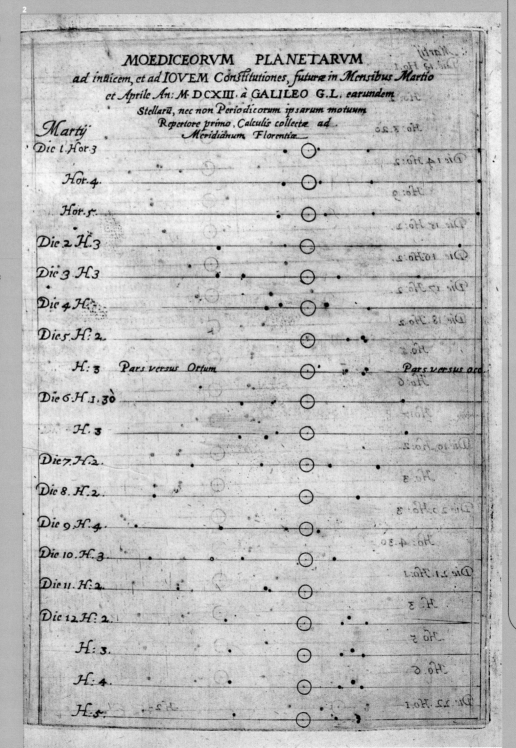

Such things happened to the *Galileo* spacecraft during its closest approaches to Jupiter in the early twenty-first century. The raging radiation would crash its software and shut down the computers. I know *Pegasus* is tougher, but it's also carrying a crew, and *we* only shut down once. There's no re-booting procedure for a human.

Tom's been putting us all through rigorous simulations of the aerobraking manoeuvre, practising every disaster scenario that Control could upload to us. We know just what to do if an engine fails, if we hit storm turbulence, if our magnetoshield loses power, or if we collide with debris in Jupiter's ring as we cross the Van Allen belts. Whether we can do any of these things when we're pulling 9 Gs, I don't know. No-one's really discussed that. But the last simulation is now over, and stuff is being stowed away to protect it from the Gs. The next one is for real, and there's no second chance half a billion miles from Earth.

⭐ **Tom Kirby, Commander**
411 days, 19 hours, 55 minutes (Jupiter orbit)

Everyone is completely exhausted. With the exception of Yvan, who had to endure a punishing EVA on Venus, none of us has

Kirby – exhausted after the Jupiter aerobrake.

been through anything like that since we left Earth. But at least we made it.

Pegasus has now slowed to 11 km a second and is on an elliptical orbit around Jupiter. Our course will intersect with Io in another 10 hours. The ship performed well during the aerobrake, but Zoë and John didn't fare so well – both of them passed out as the G-forces peaked. They are conscious again now, but a little worse for wear. John's got some bruising from ruptured blood vessels in his legs and back, but nothing serious.

We hit the bow shock 4.75 days from Jupiter's cloud tops, with 9 million km still to go. It's a dynamic boundary that ebbs and flows as the pressure of the solar wind shifts, and we crossed and recrossed it for about 24 hours. Our magnetoshield was powered up during this transition and has performed well so far.

We synchronized our approach to Jupiter with the planet's rotation. This kept the ship in a weak area in Jupiter's magnetic field, where the radiation is less intense. There was little left to do now but wait. All the same, we scanned checklists anxiously as we approached the Van Allen belts. An abrupt rise in radiation levels and a bright glow from the magnetoshield signalled our arrival. We crossed the belts as fast as possible, holding back the engine burn that was to put us into orbit. The burn then slowed us from 30 km per second to 22 km in just six minutes, before the main engine was hurriedly shut down and the blast doors closed, with just three minutes to go before we hit the upper atmosphere.

Braking started with heavy vibrations in the truss and a growing roar. After nine months of relative quiet on *Pegasus*, the noise was shocking – it sounded like the craft was in agony. The aeroshield temperature shot up, and so did the Gs. My G-suit began to constrict around my legs, and the mounting force compressed my chest and flattened my face. I was trying to read out the Gs, but the strain of speaking became unbearable when I got to 8 Gs. I inhaled deeply and blew back against my closed mouth, forcing pressure up into my eye sockets to try and preserve their shape and my vision.

We were on the nightside of Jupiter, and the only light was coming from the glowing aeroshield. Cameras outside the habitation module caught the fierce orange glow of plasma popping and flashing off the disc at the other end of the ship. I'd never seen *Pegasus* looking so animated.

Our discomfort peaked at 11 Gs, but this quickly dropped back to about 4 Gs. We sustained 4 Gs for more than 40 minutes before

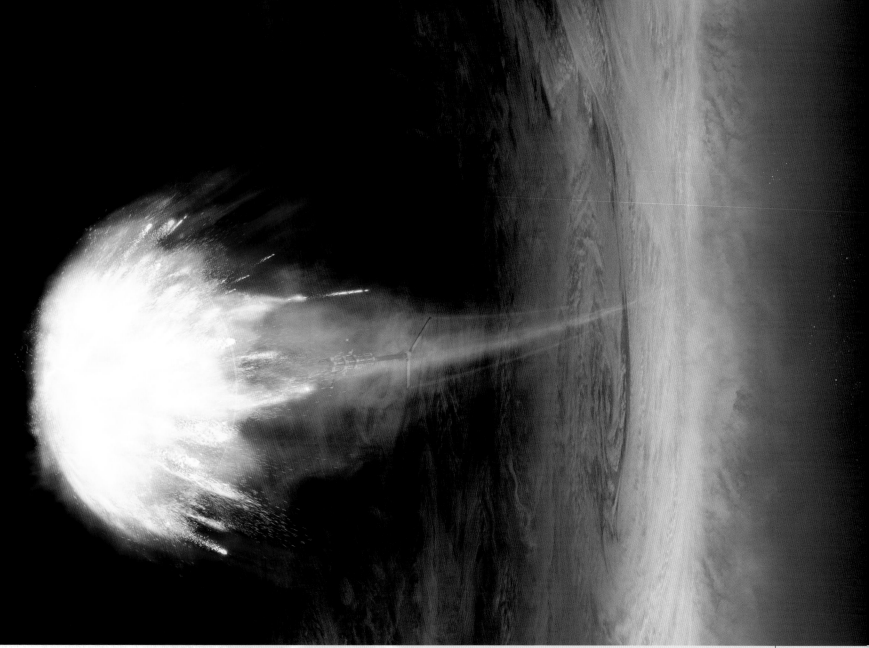

Pegasus **tears through the cloud tops of Jupiter in a brutal aerobraking manoeuvre.**

skipping out of the atmosphere on the other side of Jupiter. The Gs instantly dropped to zero. The elephant I'd felt sitting my chest suddenly jumped off, and the straps on our seats floated upwards again. This sensation was accompanied by a sudden silence as the turbulence and vibrations stopped abruptly. Yvan opened the shutters. It was dark outside – just a slim, crescent-shaped Jupiter in the blackness. All of a sudden, a bright flash of light whizzed past. At first I thought it was a bit of *Pegasus* breaking off, but Yvan said 'falling star'. Jupiter is continually pulling in debris at high speed, and most of it burns up just like we nearly had. We'd ridden our own meteor through the clouds of Jupiter and survived.

Io here we come.

Right: The crew prepare for the Jupiter aerobrake.

Inside the Giant

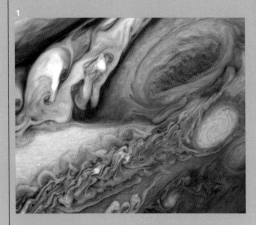

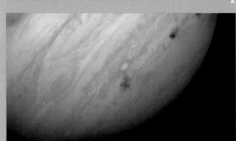

If aliens from another star system ever try to detect planets in our solar system, they'll spot Jupiter first. Jupiter is so massive that it tugs the Sun this way and that as it orbits. It is more than twice the weight of all the other planets put together.

Jupiter consists mostly of hydrogen and helium. The swirling colours we see in the clouds come from a cocktail of other chemicals, such as ammonia, methane and water ice. The motions of these clouds probably extend down for thousands of kilometres.

As in any atmosphere, air rises and falls in cells. On Jupiter the cells are stretched into wide bands by the planet's rapid rotation, giving the planet its striped appearance. The pale stripes ('zones') are the highest clouds, formed by rising air; darker stripes ('belts') are lower, descending air currents. Jupiter's rapid rotation also makes the planet bulge in the middle, giving an equatorial diameter 9000 km (5600 miles) greater than the distance between the poles.

There are gaps in Jupiter's clouds – patches of hot, dry air through which we can peer to about 50 km (30 miles) deep.

Down there the air is comfortably warm – about room temperature – a stark contrast to the freezing ammonia cloud tops, where the temperature drops to −150 ºC (−238 ºF). Deeper still, friction between rising and falling air currents causes a massive build up of electrical charge, triggering bolts of lighting a thousand times brighter than any on Earth.

The most prominent features of the atmosphere are the long-lived storms, and chief among these is the Great Red Spot, a 26,000-km (16,000-mile) wide cyclone that's been seen on and off since the seventeenth century. It rises 30 km (19

miles) above its surroundings and rotates once every seven days. Its distinctive colour has been the subject of much debate – some scientists think the colour comes from a chemical called phosphine; others blame carbon-based chemicals.

Jupiter's giant magnetic field is apparently generated by its inner ocean of metallic liquid hydrogen, which is constantly being churned by the planet's fast rotation. The field forms a gigantic bubble that is blown into the shape of a comet by the solar wind. On the sunward side, the bubble extends some 3 million km (2 million miles) into space; the 'tail' stretches away from the Sun and reaches as far as Saturn's orbit, 800 million km (500 million miles) away.

In 1994, less than 17 months before the *Galileo* spacecraft reached Jupiter, comet Shoemaker–Levy 9 ploughed into the far side of the planet in a stream of 16 fragments. Astronomers all over the world watched in wonder as Earth-sized impact scars rotated into view over the next few hours. As each fragment buried itself, explosions of heat and light threw up tonnes of deep material, revealing a range of chemicals that had never before been seen on Jupiter.

1. Jupiter's Great Red Spot – a giant storm larger than the entire Earth.
2. Fragments of comet Shoemaker–Levy 9 shortly before impact with Jupiter in 1994.
3. The scars in the Jovian atmosphere left by the comet fragment impacts.
4. Moment of impact as part of SL9 smashes into Jupiter – causing a bright flash of light on the nightside of the planet.

Jupiter or Bust

Missions to the Moon, Mars and Venus may have seemed ambitious in the 1960s, but all are close neighbours. Jupiter was the first planet in the outer solar system to be targeted, and reaching it required an extraordinary amount of planning, technology and luck.

First there was the largely uncharted asteroid belt to deal with. *Pioneer 10* was charged with the task of blazing a trail through. It entered the belt in July 1972, and mission controllers spent a nail-biting seven months waiting to see if their priceless little robot would end up smashed by a rock. But by February 1973, *Pioneer 10* had safely sailed through, and Jupiter beckoned.

As Jupiter loomed close, *Pioneer 10* started to pick up much stronger magnetic fields than anyone had expected. During its closest approach, at 130,000 km (81,000 miles) from Jupiter's cloud tops, the magnetic field was 10,000 times stronger than Earth's. More worryingly, powerful X-rays and gamma rays generated by particles in the magnetic field began to damage the probe's instruments. But *Pioneer 10* survived and managed to radio back details of Jupiter's chemical composition and the first close-ups of the planet and its moons.

As *Pioneer 10* coasted past Jupiter, becoming the first spacecraft to use a planet's gravity as a 'slingshot' to boost its speed, NASA was rushing to complete work on two follow-up craft. The new probes were to take advantage of a rare planetary alignment that would give them close encounters with Jupiter, Saturn, Uranus and Neptune. A competition was held to name

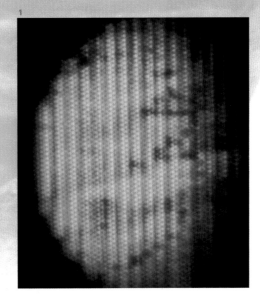

them, and the winning entry was Voyager.

Voyager 1 and *2* left Earth in autumn 1977 on what was to be a spectacularly successful mission. They reached Jupiter nearly two years later, when *Voyager 1* recorded a breathtaking time-lapse film of turbulent eddies swirling around the Great Red Spot. Back on Earth, the scientists were spellbound. Although Jupiter's intense radiation interfered with their computers, the probes sent back over 52,000 images of the planet and its moons. They discovered a thin ring around the gas giant and active volcanoes on Io before heading on to Saturn.

The Voyagers' discoveries spurred NASA on to return to Jupiter – but this time they wanted to stay. The radiation-hardened *Galileo* mission arrived at Jupiter in 1995 – releasing a probe that descended over 1000 km into the planet relaying data on temperatures, pressures, wind speeds and cloud chemistry. The main spacecraft continued to orbit the planet for over seven years – making multiple flybys of the big

Galilean moons and observing the planet's weather systems more closely. But by 2003 it was suffering the effects of radiation damage, mission planners decided to dispose of it safely – sending it on a suicide-dive into the atmosphere of Jupiter. A future mission is planned to go into orbit around the icy moons in a couple of decades.

1. The first close-up view of Jupiter taken by *Pioneer 10* – visibly strained by the extreme magnetic field the spacecraft was immersed in.
2. A better *Pioneer 10* image of Juputer.
3. Next to the red spot is a white oval that formed in 1939 and has remained intact ever since.

Lessard suits up for Io.

Mission Control monitors radiation levels on Io.

⭐ Tom Kirby, Commander
413 days, 11 hours, 12 minutes (Io orbit)

I can't believe they've sent her to Massaw – she'll be exposed to the harshest radiation on that side of Io.

We were all worried when Io came into view and Yvan spotted that our primary site, a volcano called Prometheus, had erupted and covered the landing site with fresh sulphurous ash. But we knew Io was unpredictable, and that's why we had four possible sites. Then we discovered that the second site, Ra Patera, had levels of radiation too lethal to venture into. That left Media Regio – a safe plain far from anything interesting – and the volcano Massaw Patera, which we also considered too close to the strong radiation fields. I talked it over with Zoë before filing my report to Control. She was very reluctant to accept the plains site. It has no volcanic craters, but it was the least risky.

Within two hours Control had overruled us and returned instructions to go to Massaw. They said they were confident the radiation shielding in the lander and Zoë's suit would be up to the job. Even so, they've restricted Zoë's time on the surface to just four hours.

Zoë seemed happy with the compromise and thinks she can get more than they are proposing done in the four hours. But John is so distressed about the radiation that he's hardly talking. Yvan is worried too – and that bothers me. He's the only one of us who's come close to what Zoë's about to do, during his walk on Venus in an equally cumbersome suit more than a year ago. The exhaustion he suffered back then is something we haven't forgotten. This is another asteroid moment – Control are pushing us too hard. I am going to have to do something soon for the sake of my crew.

Hermes left the relative safety of *Pegasus* a few moments ago, after we'd reached the less dangerous side of Io. The lander's rockets will propel it rapidly to within 20 km of the surface, after

which Zoë will fly low over the terrain and east towards Massaw. Good luck Zoë – all our prayers are with you.

➕ John Pearson, Flight Medic
413 days, 12 hours, 20 minutes

Hermes is on the ground at Massaw. Zoë's got a lot to accomplish in less than four hours, and we all know how hard it is to work in that suit.

Descent and landing went well. I watched *Hermes* from the cupola as it undocked, its foil body glinting in the bright sunlight on the day side of Io. The spidery little lander looked so tiny and vulnerable against the huge, blistered surface of this daunting world. Little puffs of propulsion gas flipped *Hermes* over so its five legs were facing forwards and slightly upwards, away from the moon's surface. After a brief countdown, Zoë fired a sustained burn to shed orbital speed and so send the lander accelerating towards the mottled surface below. *Hermes* soon disappeared beneath us and I returned to the command centre to follow Zoë's progress.

Hermes is a lightweight craft built for Io's low gravity (only an eighth of Earth's), but it has heavy radiation shielding, including a 10-centimetre jacket of titanium panels. It also carries its own magnetoshield. So far, radiation levels have been about 10 per cent lower than predicted, but I expected levels to rise again.

We listened to Zoë marvelling at the view as she got near to the surface. In fact, her excitement at seeing the volcanoes up close distracted her from the descent. Between descriptions of the terrain, she counted down her altitude. The feed from *Hermes'* external cameras came and went, crackling with interference, as the gentle slopes of Massaw came into view. As Zoë flew in,

The surface of Io snapped by the Galileo spacecraft. A dazzlingly bright lava fountain (left) overwhelmed the spacecraft camera – it might be as hot as 1600 °C.

clouds of dust began to blur the view. Confused by the radar altimeter readings, she cut the main engine a little too early. *Hermes* dropped the final 2 m to the surface with a jolt. As the vibrations of landing dampened, they were replaced by vibrations from the moon itself – as the tremors from an 'ioquake' reverberated across the landing site.

It took Zoë only 30 minutes to complete post-landing checks, suit up, depressurize the cabin, and emerged from the hatch. Now she's standing on the surface of Io, in her bulky suit, enveloped in her own aurora. It looks so alien and threatening down there. We are all very, very tense about the next few hours. I'm trying to concentrate on my tasks to stop a sense of panic rising in me – God knows what it's like for Zoë.

A plume of gas ejected some 100 km above the surface of Io.

⊞ John Pearson, Flight Medic
413 days, 15 hours, 6 minutes

It has been a harrowing few hours. I had to pull the plug on Zoë's EVA. I shouldn't have let her out in the first place. If she doesn't get back, I'll blame myself for the rest of my life.

Twenty minutes into the EVA, we realized she was falling behind. She was still deploying equipment at the landing site, but she should have begun her 1-kilometre walk to the edge of Massaw's volcanic crater – the highlight of the EVA. It was when she was setting up a laser to measure tidal bulges in Io's crust that I noticed her struggling. Her heart rate was over 130. I asked Tom to tell her to relax, emphasizing that she shouldn't push herself to do everything on the itinerary. At this point, the radiation sensors on her suit still showed lower levels than we'd expected.

We were all in the command centre, glued to the camera feeds and willing Zoë on. The first sign of a serious problem came from the screens. Pixels started to drop out, and the images became intensely speckled as a blast of charged particles bombarded the cameras on board *Hermes* and soon afterwards on her suit. The signal from the suit's radiation sensors was breaking up, so I asked Tom if Zoë could resend the data. When the signal came through I was alarmed – if the numbers were right, the suit's magnetoshield had a breach. I suggested she turn up the power. Zoë had spotted the problem too, but insisted she could go on. Her heart rate was at a steady 150 for the next hour, and radiation levels fluctuated wildly as magnetic field lines swept across the landing site. Her suit was stiff and cumbersome, and the trolley that was designed to help her kept getting caught in the 'snow' drifts of sulphur crystals coating the ground.

Zoë was already halfway to the Massaw crater when Yvan noticed that the increased power being used to boost the suit's magnetoshield was sapping the life-support system. She would have to come back, and that's when I decided to pull the plug.

As I write, she's still somewhere between the lander and the crater, and night is falling. Soon it will be dark. She's exhausted, and getting back will take an immense effort. I know she'll be disappointed that I've cut the mission short, but I just hope she lives long enough to be angry with me.

☒ Yvan Grigorev, Flight Engineer
413 days, 18 hours, 30 minutes

Almost all the power in Zoë's suit was used up before we got confirmation that she was back inside *Hermes* with the hatch sealed. The walk back was torture – she kept stopping and going silent on us. At one point we were so desperate to get her moving again that I blasted some of her least favourite Russian rock music into her headset. Night had fallen by the time she reached *Hermes*, and although the twilight cast by Jupiter was enough to stop her stumbling on the lava flows, we had to turn on the lander's floodlights to guide her back.

Then, inside the lander, we lost visual and audible contact with her. John was still getting medical data, and we knew she was alive but possibly unconscious – certainly not in a fit state to pilot *Hermes* back into orbit. There was a danger she might asphyxiate inside the suit if the life-support system was completely drained of power and she was still sealed inside it.

To get her back as fast as possible, I launched *Hermes* remotely before we were in the right part of our orbit to rendezvous with it; Tom moved *Pegasus* into a lower, faster orbit to catch up with *Hermes*, rather than holding Zoë on the surface until we passed overhead. We made the rendezvous in record time, dropping *Pegasus* faster than we should have done. With the two spacecraft docked, I left Tom at the controls and pulled myself through the airlock and into *Hermes*. Zoë was wedged against the floor of the lower deck, inside her suit but still breathing.

She is resting now, with John looking after her. As soon as I had the chance, I took care of the suit. I know we were meant to hold on to it, since it was covered in priceless bits of Io, but I told Control it was too radioactive and threw it out of the airlock. It was a rubbish design and I blame it entirely for the failure of the mission. It almost killed Zoë.

💬 Zoë Lessard, Mission Scientist
414 days, 16 hours, 29 minutes

It's hard to write this. My whole memory of the mission is tainted by failure – failure to complete the EVA; failure to reach the Massaw crater; failure to get myself back into orbit; and failure even to bring back a single rock sample. And that's what hurts the most – all that training, studying and planning, and I came back empty-handed. I also scared my friends, who risked their lives and the whole mission to rescue me. Yes, I got the experiments laid out, but robots could have done that. I went to bring back rocks, to collect those exotic, primeval lavas that only an expert human eye can pick out. They were all around me. I'd even put them in the sample box. It was sitting there on the surface as I staggered back to *Hermes*; all I needed to do was pick it up. It sounds so easy now, but it was all I could do to pick up each foot to get myself back into the lander.

Control have been very understanding, congratulating me on what I accomplished, but all I can think of is what I didn't do. Tom says the schedule was too ambitious, that I did everything I could. Yvan has been very good about it too. He says it was the suit's fault, not mine. But I'd trained hard in that suit in endless simulations. The simple fact is that I didn't pace myself. I tired too quickly and failed to complete the EVA. I let everyone down.

From orbit it had all looked so easy, and I was so pumped up. Io looked spectacular. As I flew towards Massaw, the Sun was behind me, casting long shadows off the highest peaks. Beyond the landing site, the horizon was disappearing into the night, with only the highest peaks of mountains catching the last rays of evening sunlight. They looked like they were floating, their bases already obscured by the darkness.

The little moon was giving me a few navigational problems. Irregular concentrations of mass in the crust were changing the strength of Io's gravity, throwing my planned orbit off course. Although we'd expected it, the lumpy gravity was confusing the computer. This sure was a grumpy hunk of rock!

Approaching Massaw, I saw its spectacular vertical cliffs, which are more than 2 km tall. We'd short-listed this landing site because the volcano is thought to be dormant. But, eruptions on Io are not fully understood, and the only way to check that the site was safe was to use a heat-sensitive camera on *Hermes* to locate a cool zone. I needed to land within walking distance of the crater.

The dust caused problems on landing. *Hermes'* engines blew up a cloud of it while I was still 30 m up, and visibility deteriorated. I strained to maintain eye contact with the ground so I could see how fast I was moving. But the visibility just got worse as I got lower, and, misreading the altimeter, I cut the engine slightly too soon, when I was a couple of metres off the ground. It didn't make for a very smooth landing – *Hermes* dropped and the jolt startled me. I braced myself, half expecting the lander to topple over.

But the ground held. *Hermes* had settled at a slight angle, about 4 degrees. This was within safety margins, and Tom agreed I could stay where I was. As soon as the oscillations of landing stopped, I became aware of a faint tremor – the ground was shaking. This 'ioquake' wasn't like an earthquake or an aftershock on Earth – it just went on an on, without end.

sky, and opposite it Jupiter's colossal face as big as 40 full moons seen from Earth. It hung there still, silent and oppressive. But there was no time to stand and stare; I had work to do.

I began collecting rocks. Everything was covered in yellow or brick-red dust – sulphurous ash spewed out by Io's volcanoes. It was tiring to walk on, and it clogged the wheels of my instrument trolley, which quickly became a burden to haul behind me. In some places the ground was clear of dust and the underlying silicate lava was exposed. It was rough and glassy, and crunched underfoot. There were also patches of older crusty sulphur dioxide snow, which was firmer and easier to walk on. At first I moved very cautiously in the low gravity as I was anxious not to fall and damage the suit. But as my confidence grew, I became bolder and perhaps a little reckless, and I fell over a few times as I hurried to collect samples and set up the science

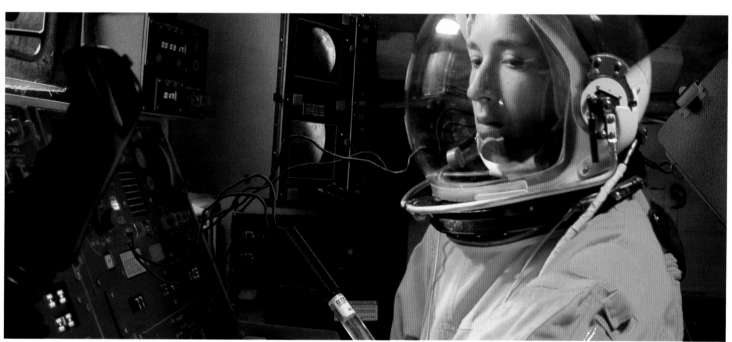

Lessard begins her descent in *Hermes* to the surface of Io.

Within 30 minutes I'd depressurized the cabin and broken the seal on the hatch door. Cameras outside the lander relayed pictures to my helmet's monitor, guiding me backwards down the rungs of the ladder. I felt a bit wobbly, unbalanced by the weak ⅙th gravity and anxious not to slip and make a fool of myself. Brown dust thrown up by the landing had stained the outside of *Hermes* and covered the footpad. The dust was as fine as talc, and the lander's feet had sunk perhaps 30 cm into it. For a moment I was worried I'd sink to my waist in it, but the ground was solid. I let go of *Hermes* and took a step. I was on Io – my dream had come true. I took in the view: a diminished Sun hanging low in the jet-black

instruments. Tom had to remind me to take it easy a few times, but I remember thinking he was being overprotective.

Time was slipping by quickly, and it was a push to get everything done. But things seemed to be going well, though my suit's magnetoshield wasn't performing properly and Tom advised me to turn the power up. When I'd finished setting up the science gear, I set off on the long walk to the edge of the Massaw crater. At last I had some time to look at the scenery. Far on the horizon, volcanic eruptions burst into the black sky, throwing up vast plumes of dust that glowed as they scattered the sunlight.

Lessard pauses by a frozen lava column on the surface of Io.

I was lost in thought when Tom interrupted me and told me to get back to *Hermes*. I was shocked. The highlight of my visit was still to come, and I didn't want to turn back. I pretended not to hear him and continued trudging through the dust, dragging the instrument trolley behind. But the truth was I was exhausted and dehydrated after four hours of relentless exertion in the bulky spacesuit.

It was when I stopped to talk to Tom that I started feeling faint. It brought me to my senses. I couldn't afford to pass out here – I could die standing in my suit. On top of this, the Sun was going down and the light was failing. I turned back and started trudging towards *Hermes*, as the sun set.

Jupiter was behind me now, casting a brown light across the ground. The temperature was falling, and sparkling crystals were forming on my suit. I remember thinking how pretty it was. I just wanted to stop right there, rest my aching muscles, and watch it. Silly bitch.

The next thing I remember was being jerked out of my trance by Yvan's dreadful rock music and Tom's voice, ordering me to return to *Hermes*. I had to get moving – and fast. Beautiful though it was, this place was now killing me.

I don't know how I got through the exhausting last half a kilometre to *Hermes*. The landscape was much darker, and I could see little but the lander's floodlight in the distance. I headed towards it, dimly aware of a wavy aurora that was spreading across Io's southern sky. That's the last memory of my botched mission to Io. Now I can't help wishing I really had died there.

Nina Sulman, Mission Scientist
416 days, 8 hours, 10 minutes

We're just one day away from our rendezvous with the *Shackleton* probe, which is now back in orbit around Europa. It's another chance of finding evidence that we're not alone in the solar system, and it should be an exciting time, but the mood in here is very depressed. The burden of Zoë's aborted EVA on Io is weighing heavily on all of us. Yvan is angry at Control. John is just very quiet and looks drained. Tom feels like he let Zoë down, and Zoë feels like she's let everyone down. And there seems to be nothing I can say that will change how any of them feel.

Lessard shows her strain on the EVA.

If there was ever a good time on this mission to make a great discovery, then this is it. But could there really be life on Europa? I keep going through the logic in my head. We are pretty sure the planet hides a liquid ocean many kilometres below its icy crust. This is partly because the moon has so few impact craters, which suggests that liquid water fills them in and freezes over. There are also canal-like streaks crisscrossing the surface, and rafts of ice that seem to have drifted before refreezing. And then there's Europa's magnetic field, which hints at something deep inside the planet that conducts electricity – maybe brine.

Everything points to a hidden ocean, but if it's buried under 20 km of ice, any life there might as well be in another galaxy. Despite centuries of drilling for oil on Earth, we still don't have the technology to drill that deep into Europa. However, meteors might have done the job for us. In the floors of some impact craters, the icy crust is probably thinner. And there are tantalizing patches of colour in Europa's ice – stains that might come from upwellings of warm water carrying organic material from deep below.

We already know that *Shackleton* succeeded in collecting 22 ice samples, the deepest from almost 100 m. These have been melted and filtered to collect any solid material, and we all want to know what that solid material is. But that discovery is still to come. For now, as we approach Europa, there are more mundane things to worry about. New supplies of fuel and food await us here, and a fresh assortment of food parcels to squabble over might lift our spirits.

Overleaf: Alone on the Ioan surface shortly before sunset and Lessard is still a long way from *Hermes*. Jupiter in the background appears 'upside-down' when viewed from the southern hemisphere of Io.

☒ **Yvan Grigorev, Flight Engineer**
421 days, 11 hours, 22 minutes

Europa is a weird-looking place. It looks likes a frozen sea and reminds me of a winter's day on the shores of St Petersburg. But compared to Jupiter and Io, it's a monotonous surface, and the views aren't helping to lift our spirits. In the last few days, we've

Two days ago, Tom and I dispatched a high-speed probe into Jupiter's atmosphere to study the Great Red Spot – a storm that's been raging for hundreds of years. Although the probe's official name is *Juno* (after the wife of Jupiter in Roman mythology), we've christened it *Chuck* after General Chuck Yeager, the first man to break the sound barrier back on Earth. After *Chuck* entered the atmosphere and jettisoned its heat shield, a nuclear ramjet engine kicked in, propelling it deep into Jupiter at hypersonic speeds. The probe should survive up to 100 days descending up to 1000 km into the planet as it soars through Jovian skies.

Juno is released from *Pegasus*, still cocooned in its shield for the violent descent into the clouds.

all been wondering what we're doing here, so far from home. The setbacks on Io, the list of repairs, and the high levels of radiation are sapping our energy.

We know things will calm down on the next leg of the voyage, as we head out to Saturn. But for the first time, the prospect of travelling even further form Earth seems overambitious and maybe even futile.

So far it's working like a dream, but our celebrations are muted. *Chuck* is travelling through clouds of ammonia and water that might just harbour life – like the floating jellyfish that some scientists have wildly speculated about. But unless we capture them on camera, we'll never know. *Chuck* doesn't carry any exobiology experiments – it's only designed to study the weather.

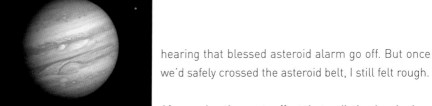

John Pearson, Flight Medic
429 days, 17 hours, 6 minutes

I'm sure the others have noticed there's something wrong. I've not been myself since we reached Jupiter, and the strain of aerobraking took a greater toll on me than on the others. Not only did I lose consciousness, but the bruising I suffered was

hearing that blessed asteroid alarm go off. But once we'd safely crossed the asteroid belt, I still felt rough.

After seeing the nasty effect that radiation has had on me, it was hard to watch Zoë risking her health on Io. I have kept a close eye on her since, and looked for signs of radiation sickness – nausea, hair loss, and so on. But she seems fighting fit, at least physically. The 'failure', as she sees it, still weighs heavy on her. I've prescribed some amphetamines and an exercise regime to lift her mood. I know the drugs will get into

The nuclear powered ramjet *Juno* unfurls its wings and heads towards Jupiter's immense red storm system.

worse than it should have been after a brief bout of 11 Gs. I felt lousy for days, and I'm still in a lot of pain.

The chief flight surgeon, Clare Granier, knows about my condition – I had to confide in someone. I e-mailed her the results of the blood test I'd done on myself, partly because I wanted a second opinion. I'd suspected lymphoma since I started waking up in a cold sweat not long after we flew through the solar corona. At first I tried to blame it on the weeks spent living with the fear of

the drinking water, but they could do us all a bit of good! I wish I could cure myself so easily, but the medicines I need would poison the water – chemotherapy isn't something to share with friends. I can only hope that a positive outlook and a good diet will beat it.

I'm not sure how much longer I can keep it from the others. I'm losing weight and looking paler. The pain is making me grumpy. I'm sure Tom already suspects something – it doesn't take a medic to spot that I'm not well. But as a medic, I know I am dying.

Overleaf: Pegasus in orbit around Europa – still in the intense radiation belts of Jupiter – the poles of the magnetic shield glow with their own aurora.

Moons of Ice, Fire and Rock

moons, including the four big Galilean satellites, which rival the other planets in size and orbit Jupiter's equator in circles. Farther out are six smaller moons, with elliptical orbits at an angle to Jupiter's equator. The rest are even farther and are probably captured asteroids, some orbiting Jupiter in the opposite direction to its rotation.

Counting the moons of Jupiter has become quite a pastime in recent years. Only four were known until the twentieth century, but then the number began to creep up, reaching the teens by the year 2000. In 2003 another 45 were added when astronomers used the world's the biggest digital camera to photograph the Jupiter system from a mountain-top observatory in Hawaii. Today the figure stands at 60.

Jupiter's moons can be divided into several groups. There are eight inner

The innermost Galilean moon is Io, which is about the size of our Moon. Even before the Voyager probes photographed Io, astronomers had predicted that Jupiter's powerful tidal pull might have melted this moon's interior. Sure enough, Io turned out to be the most geologically active body in the solar system, with more than 100 volcanoes hurling out vast fountains of material. Most of Io's volcanoes show signs of warmth, so it is difficult to classify any as truly dormant. The craters contain molten sulphur at around 200 °C (400 °F) in the centre, and they constantly change colour as the temperature varies. Outside the craters, the temperature falls sharply to about −150 °C (−240 °F), Io's normal surface temperature.

Io is second only to Mars for its grand landscapes, with mountains that seem to defy Ionian gravity and rise to almost twice the height of Everest on a crust that shouldn't support them. Io also has a very weak atmosphere, perhaps formed from sulphur dioxide gas emitted by the volcanoes.

Europa is a little further out than Io and slightly smaller, yet it could hardly look more different, being completely covered in water ice. Some scientists think a global ocean of liquid water, perhaps 100 km (62 miles) deep, might lie below this frozen crust. Crater shapes suggest the ice is just 25 km (16 miles) thick, and patterns in the ice suggest that liquid water has seeped onto the surface in the recent past. Fluctuations in Europa's magnetic field lend support to the theory, hinting that any ocean might consist of salt water. And

places their frozen remains might be close enough to the surface to sample.

The next Galilean moon is Ganymede – the largest moon in the solar system. It's bigger than the planets Mercury and Pluto and only slightly smaller than Mars. Even at this size, however, it should have lost all its heat, but *Galileo* detected a magnetic field, suggesting the moon is still warm inside. Some scientists think it might have a partially molten iron core, kept hot by tidal heating from Jupiter and the other moons.

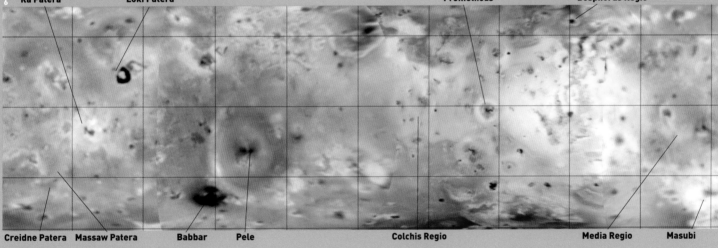

Ra Patera   Loki Patera   Prometheus   Bosphorus Regio

Creidne Patera   Massaw Patera   Babbar   Pele   Colchis Regio   Media Regio   Masubi

some astronomers think the ocean might also contain organic compounds from asteroid impacts and sulphur that has drifted across from Io's volcanoes.

Such an environment, then, with liquid water and organic compounds, provides the key ingredients for life. There are even yellow-brown patches in the ice that have been interpreted as stains made by bacteria-like organisms flourishing in the water. Their spectral signatures are not dissimilar to those of bacteria that live in extreme environments on Earth, and in

Callisto, the outermost Galilean moon, is also larger than Mercury. This moon has a very dark surface covered with an incredible number of craters – a sign that the exterior is geologically inactive and hasn't changed for billions of years. Some of the craters run in straight lines across the surface. These may be the scars of comets torn into fragments by Jupiter's gravity, as happened with Shoemaker-Levy 9. Like Europa, Callisto shows some signs of a subsurface ocean, but it has not been as closely studied.

1. The jumbled surface of Europa suggests that these blocks are floating on a liquid interior.
2. A global view of the Jovian moon Ganymede's frosty polar caps and bright, grooved terrain.
3. The Zal Patera region of Io – these mountains are between 2 and 4 km (1 and 2 ½ miles) high.
4. An active lava flow on Io over 60 km (37 miles) long.
5. Bright scars on a darker surface testify to a long history of impacts on Jupiter's moon Callisto.
6. A false-colour map of Io.
7. From top to bottom: Io, Europa, Ganymede and Callisto shown to scale against the Great Red Spot.

Of Rings and Moons

SATURN

⊠ DIAMETER: **120,536 KM (74,875 MILES)** ⊠ MASS (EARTH = 1): **95** ⊠ GRAVITY AT CLOUD TOPS (EARTH = 1): **1.19**

⊠ CLOUD-TOP TEMPERATURE: **−180 ºC (−292 ºF)** ⊠ DAYLENGTH (ROTATION PERIOD): **10.21 EARTH HOURS**

⊠ YEAR: **29.4 EARTH-YEARS** ⊠ DISTANCE FROM SUN: **1427 MILLION KM (887 MILLION MILES)** ⊠ RINGS: **7** ⊠ MOONS: **30** ⊠

After the near-disaster on Io, there was a vociferous backlash in the press. It was as if our detractors had been waiting for just such an event to prove them right. Out came all the negative arguments about the cost and safety of the project, and everyone seemed to forget our achievements. It affected the morale of the ground crew pretty badly, but there was no point searching for scapegoats. There were five astronauts still up there depending on us, and they had no choice but to continue. So, like them, we had to put the past behind us and move on. There would be other chances to explore Io – perhaps not in my lifetime – but we would return.

Even during her 'failed' mission, Zoë installed more equipment on Io than any robot could have done, and it continues working to this day, monitoring the endless tremors of Io's crust and radiation fluxes from Jupiter. Mercifully, the burst of radiation recorded in her suit turned out to be a faulty reading. Her radiation exposure had stayed within safety limits, and the dizzy spells she suffered were caused by nothing more than exhaustion. She was soon back to full health.

The ramjet probe that entered Jupiter survived for 130 days, a month longer than we'd anticipated. It recorded wind speeds, temperatures, pressure and atmospheric chemistry to great depths. It sent daily weather reports from the largest weather systems in the solar system from the Great Red Spot to the giant jet streams of the poles.

The successful retrieval of water ice from Europa reopened the debate about whether this tiny moon could support life. Nina began her careful examination of the sample, and all of us hoped for something tangible to tell the press. Predictably, the media had focused on the search for life, and so far we'd found nothing. To make matters worse, rumours were circulating at Control about one of the crew being sick. With all the bad press and the talk of death, a discovery of life in the oceans of Europa was badly needed.

Jupiter was now behind us, and the alien storms of another gas giant were beckoning. But a year of interplanetary travel lay between *Pegasus* and the Saturn system – a year of repairs, analysis, planning and studying. As it turned out, it was also the last year for one of the team.

Alex Lloyd
Chief Scientist, *Pegasus* Mission

◉ Zoë Lessard, Mission Scientist
Mission elapsed time: 501 days, 1 hour, 8 minutes

- -

Despite what some people are saying, I am not clinically depressed, and I am still pulling my weight. I may be obsessing a bit, but that's because I'm finding sleep difficult and have too much time to think. Tom keeps trying to persuade me there was nothing I could have done about Io, but I just find his long talks irritating. I feel better pounding the treadmill and working things out myself.

I feel abandoned by Control. They don't seem to want any debrief or analysis of what went wrong on Io. Just move on, they say. But I can't. Two months on from Jupiter, and Io still haunts my sleep. Every time I think of the mistakes I made on the surface, my stomach lurches. And the comments from Earth haven't helped. I know Control try to screen things, but they couldn't entirely hide

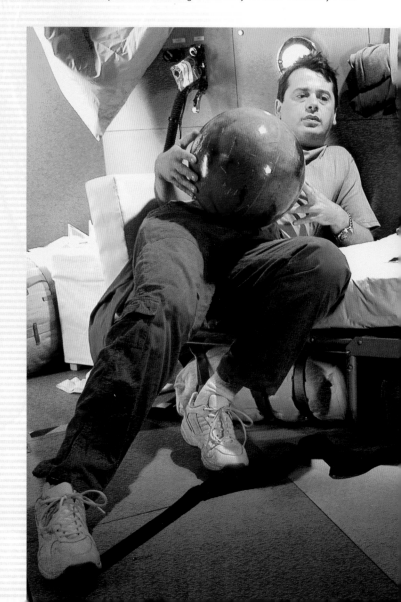

Lessard in her sleeping quarters – ponders her brush with death on Io.

the controversy that erupted back home after Io. And it was all so unfair and misinformed and one-sided. Maybe being cooped up in a metal box millions of miles from home makes you more sensitive, but what we need is support, not sniping criticism.

Grigorev tries to cheer up Lessard.

Tom Kirby, Commander
501 days, 10 hours, 22 minutes

It was with great relief that we finally ignited our engines and blasted away from Jupiter. Its potent magnetic field plagued every moment we spent within it, shutting down systems on *Pegasus*, hampering communications with Earth, and almost killing Zoë on Io. The stress was not good for the crew. But now we've returned to the relative calm of interplanetary flight, and Jupiter is just a memory. All we can see of it is a thin crescent, and even that is increasingly hard to see against the glare of the Sun. I hope its passing helps Zoë get over things. We've talked endlessly about what happened, but she's still very down and spends a lot of time alone. We'll have to make allowances for her.

Saturn is already clearly visible through the telescope, though it's still eight months away. But it's not Saturn that dominates my thoughts – John's health is my biggest worry. I'd suspected for some time that he was ill, but our recent chat has revealed it's much worse than I thought. Lymphoma is a kind of cancer, and it's not something we can cope with effectively on the flight. We're not carrying drugs for cancer, and even if we were, they'd get into the water system and make us all sick. Even strong painkillers pollute the recycled water. John has been in a lot of pain lately, and he tells me the condition shows no signs of improving.

I'm at a loss as to what to do – there's no procedure for tackling cancer. We've just passed the 500-day mark, and the earliest I could get us on a return trajectory would be after Saturn. At best it would take another 500 days to get home, and I don't know if John can survive that long. The grim truth is that I am looking at the possibility of losing him.

John broke the news to the others during a meal break. They'd all guessed something was up. We discussed it for a couple of hours, and we've urged John to start taking painkillers. Yvan will try to isolate part of the water system so the drugs won't get into our drinking water. We've reduced John's duties, giving him more time to rest. He's very optimistic about things and keeps talking about the possibility of remission and how great it will be to see Saturn close up.

Disappearing Planets

Early astronomers didn't know that planets could have rings, so they were understandably confused by Saturn. The Italian astronomer Galileo was the first person to view Saturn's rings through a telescope, in 1610. He interpreted them as companion planets that were almost touching Saturn, but two years later they seemed to vanish as Saturn passed through the point in its 15-year cycle where the rings are seen edge-on from Earth. Galileo was totally stumped by their disappearance. When they reappeared a couple of years later, his best explanation was that Saturn had two great arms, or handles.

Improvements in optics over the following half-century revealed the true nature of the rings to the Dutch physicist and astronomer Christian Huygens. His brother Constantijn was a master craftsman and made exceptionally clear lenses up to 20 cm (8 inches) wide from Dutch mirror glass. Christian used these in his telescopes to make accurate drawings of Saturn in 1655, when the rings were again seen almost edge-on from Earth. But it was not until 10 years later that the astronomical community accepted that Saturn's two 'handles' were in fact a single ring. Another 10 years elapsed before the Italian astronomer Gian Domenico Cassini resolved enough detail to see that the ring was divided into two main bands. The gap between them – the Cassini division – was named after him.

Christian Huygens went on to discover Saturn's largest moon, Titan, with the same telescope he had used to sketch the rings. In his book *Cosmotheoros*, he speculated that rational creatures, perhaps even astronomers, might live on Titan. Huygens thought there was life throughout the solar

system, and he fantasized about space flights to visit our neighbours. But Titan and Saturn were so far away that he felt we'd never visit them. Ironically, a European Space Agency probe bearing his name and carrying his signature is due to land on the moon in early 2005.

1. Portrait of Christian Huygens (1629–95), Dutch astronomer and physicist. Huygens first explained the rings of Saturn, and discovered its moon Titan.
2. Saturn's rings – historical artwork by Christian Huygens.
3. Early drawings of Saturn by various astronomers spanning the period 1610 to 1650.

Lord of the Rings

Saturn is one of the four 'gas giant' planets of the outer solar system and is most famous for its immense ring system. At 120,500 km (74,875 miles) in diameter, it is nine times wider than Earth. Like Jupiter, it has a deep, turbulent atmosphere and consists mostly of hydrogen. Despite its great size, Saturn rotates quickly, completing a 'day' in just 10.5 hours. The centrifugal force generated by its rapid spin pushes the equator out in a 12,000-km (7,500-mile) bulge, making Saturn the most oval planet in the solar system.

A haze of ammonia crystals in the upper atmosphere gives Saturn a more uniform, butterscotch appearance than Jupiter.

Beneath this hazy layer is a region of ammonia and hydrogen sulphide crystals, with blue clouds of water ice and water vapour deeper still. Although far bigger than Earth, Saturn has weaker gravity, which means the planet must have a very low density. In fact, Saturn is less dense than liquid water and would float in Earth's oceans if they were big enough. The atmosphere is probably very deep, slowly turning from gas to liquid as the pressure rises towards the centre. In the centre there is probably a hot core of rock perhaps twice the size of Earth.

At an average distance of 1.43 billion km (887 million miles) from the Sun,

Saturn takes more than 29 Earth-years to complete one orbit. It is tilted at 27 degrees, which means that the northern and southern hemispheres alternately lean towards and away from the Sun during a Saturnian year. This produces winters and summers lasting about 15 years.

Detailed observations have revealed huge storms that appear as pale ovals in Saturn's atmosphere. They are thought to be caused by convection currents that carry ammonia and water vapour to higher, colder altitudes, where the water freezes to form icy white clouds that are blown around the planet by the prevailing wind. The strongest winds (stronger than on Jupiter) are found at the equator and can reach 1600 km/h (1000 mph) in an easterly direction.

1. Saturn and two of its moons – Tethys (above) and Dione.
2. An artist's impression of a shepherd moon.
3. A true-colour picture of Saturn taken by *Voyager 2*. Three moons are seen to the left – Tethys (closest to the planet), Dione and Rhea.

✚ John Pearson, Flight Medic
502 days, 12 hours, 16 minutes

It was an immense release to share my secret with the others. I'm sure they'd suspected things weren't right. I'd been neglecting my duties for the last couple of months, and the stress of my illness was tainted with guilt at letting the others down.

As principal medical doctor on the flight, I've always felt responsible for the mental and physical wellbeing of my friends. I've nurtured them through many traumas, and my announcement felt like another trauma that I had to help them deal with. I guess their welfare is a welcome distraction from my cancer. There's still a long way to go on this mission, and I'll never stop worrying about the crew – but I have to accept that I might not be around much longer to watch over them. If I die, Nina will have to take over my role.

I never know quite how I'm going to feel each day. Sometimes, when I'm weak and washed out, even the one-half gravity of the sleep modules feels exhausting, and I prefer to rest floating in the weightless parts of *Pegasus*. Other days I feel strong enough to carry out a full programme of activities. The shaving rash I'd been getting intermittently has become a permanent feature, and my night fevers are more regular. Pains in my neck and armpits have been growing, and it's a relief to be able to take painkillers.

I know it's a cliché, but I can't help thinking why me? The rigorous scans, tests and preventative surgery we went through on Earth would have picked up this disease if I'd had it in me before, so it must have begun during the voyage. It's impossible to prove, but I think the trigger was the flight past the Sun and that extra dose of radiation that pushed my dosimeter into amber. I'm trying to accept that the answer to my question is just bad luck.

But I'm not an unlucky person. I was lucky to be a medic, lucky to get onto the astronaut training programme, lucky to fly on a space shuttle, and lucky to visit the lunar base in the Ocean of Storms. But to fly on a grand tour of the solar system – that was the greatest luck of all. Now I've floated above the cloud tops of Venus, walked on Mars, and lived among the moons of Jupiter – it's more than anyone could hope for in a single lifetime. And if I can stay alive for eight more months, I'll have the chance to reach the rings of Saturn. There must be millions of people on Earth who'd die to be where I am right now. I am not unlucky at all.

ℹ Nina Sulman, Mission Scientist
503 days, 9 hours, 19 minutes

Work on the Europa samples occupies most of my days at the moment. The ice cores were damaged when *Shackleton* took off from the surface, and it's been hard to work out which depth each of the samples is from. But the focus of my investigation has been the opaque material buried in the ice. Spectral analysis confirms a likeness to some Earth bacteria, and results from the mass spectrometer suggest organic material. But I can find no cellular structures in the material, and without them it's hard to conclude that this ice contains any living organisms, or even a corpse. My only hope is that the final tests – where I'll try to

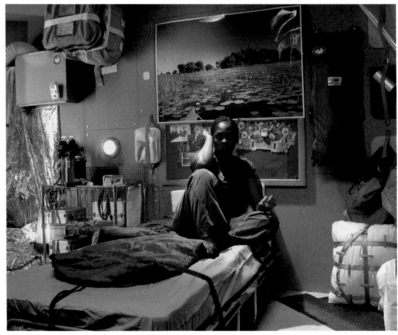

Sulman contemplates the lack of biogenic material in her Europa ice samples.

nurture any bugs in the sample to life by gently warming it – will bear some fruit.

Mission protocol dictates that I examine only a small portion of the total sample and leave the rest untouched for return to Earth. It pains me to leave so much of the ice hidden in the sample containers, knowing that they might contain the material the world is waiting for. But I can't even take a peak – if I were to open them, scientists on Earth could attribute any organic material within them to contamination. To make the most of these samples, they must be opened in specially built receiving labs on Earth.

Maintenance of *Pegasus* goes on throughout the flight. Grigorev and Sulman work on the truss close to the shield and main engines.

I don't mind admitting that, this far from home, I'm seriously tempted to unwrap a few more samples. If we were all wiped out by an asteroid impact tomorrow, the details I'm sending back now would be our only means of telling whether there's ever been life on Europa. And we risked our lives to collect these samples – perhaps Control should learn to trust me a bit more.

⊠ **Yvan Grigorev, Flight Engineer**
540 days, 17 hours, 20 minutes

Jupiter was a brutal place and *Pegasus* took quite a beating. The backlog of repairs has already consumed more than three months with most of us working flat out, and there's still plenty to

do outside the ship. I'm sure that incessant radiation has corroded parts of the hull and caused metal fatigue in the extremes of the spacecraft, outside the protection of our magnetic shield. A full survey of these parts of *Pegasus* will take up much of the rest of the journey to Saturn.

The impact analysers built into the main shield are registering just one dust grain per month as we cross the vast emptiness between Jupiter and Saturn. The last impact was on 9 July. This rate of impacts is what we'd expected, based on results from *Cassini*, the last spacecraft to come this way. Although they collide with us at over 50 km/s, each bit of dust is less than a tenth of a millionth of a metre in size. Anything bigger would cause a lot of problems. I hope it stays that way until Saturn; much as the science team would love it, the last thing I need is to fly through some unknown debris swarm.

125

⬤ Zoë Lessard, Mission Scientist
598 days, 8 hours, 00 minutes

John's bombshell has affected me in the strangest way. I feel better. I know that sounds incredibly selfish, but John's predicament has put my problems into perspective. I'd been avoiding the others, but this new crisis has pulled us together again. We are on an incredibly dangerous journey, and we are lucky to still be alive – so as John keeps saying, we have to live for the moment.

He's so excited about the prospect of reaching Saturn's rings that I sometimes forget how sick he is. It's hard not to become thrilled by his infectious enthusiasm. The bright star we've been heading towards for five months now has brightened markedly in the last few weeks and is beginning to look oval – a telltale sign of the rings. I guess this view is not dissimilar to Galileo's first view of Saturn through a telescope in 1610; he thought it had 'handles'.

⬤ Zoë Lessard, Mission Scientist
600 days, 7 hours, 21 minutes

For the last 100 days I've been running our programme of telescope observations of Saturn, examining details in the rings and watching weather systems drift across the stripy brown atmosphere. Saturn is visibly flattened by its rapid rotation – it's almost 10 per cent wider across the equator than from pole to pole, and it's the only planet in the solar system that doesn't look spherical. As with Jupiter, we're chasing Saturn along its orbit round the Sun. The rings are tilted towards the inner solar system and the planet's dark shadow looms across the rings behind, blotting out a whole section of them.

On the nightside I can just make out the elusive G and E rings, which more than double the diameter of the ring system we see from Earth. It's a truly vast carousel of debris, breathtaking even from this distance, almost five months away. The longer I watch the rings, the more bizarre they seem. We can describe their shape and movement with the laws of physics, but the shear beauty and wonder of them is beyond the language of maths or English. Sometimes all I can do is watch in awe. Soon we'll be right in them, and I know I'll be torn between the urge to just stare in wonder and the need to study them.

✦ Tom Kirby, Commander
724 days, 17 hours, 22 minutes

We've been slowing down for some weeks now, firing the rockets in reverse during sleep periods to bring down our speed so that Saturn's gravity can capture us. After Jupiter, none of us is looking forward to another uncomfortable aerobraking manoeuvre. Thankfully, Saturn's gravity is far weaker than Jupiter's, and braking into orbit should be much easier on us.

Even so, I'm worried about how John will cope. He's lost a lot of weight, and his muscle and bone strength have decreased, partly because he's been spending less time in the gravity modules. There's a danger he might break a bone during the deceleration, so we've been working hard with Control to refine the manoeuvre and pull fewer Gs.

After the first light dip into the top of Saturn's atmosphere, we'll pass the moon Pandora on our first orbit, and then Prometheus on the second orbit. More aerobraking and rocket burns will bring us closer to Saturn, and then some very delicate manoeuvres should take us into our parking orbit in the middle of the Cassini division – the prominent gap in Saturn's rings just 112,000 km from the cloud tops.

Saturn's radiation belts are feeble compared to Jupiter's – partly because the planet's magnetic field is a good deal weaker, and partly because the rings soak up radiation. So we should be relatively safe in the Cassini division, with no need for our magnetoshield. But there are other dangers. We will have to fly close to the countless rock and ice fragments that make up Saturn's rings, which means there's a high risk of collision. It will be the most complex orbital insertion we've attempted so far, but I'm looking forward to the challenge nevertheless. Simulations have gone well, and the views of the moons and ring plane will be spectacular.

John's condition, or at least his mood, seems to be improving, boosted by the prospect of seeing Saturn's rings and moons close up. Maybe he's getting better?

Pegasus rises above the Cassini Gap, 112,000 km from the cloud tops of Saturn. Dust from the ring plane is draped over the superstructure.

Saturn's Visitors

Four probes have visited Saturn, the latest arrived in 2004.

The first was *Pioneer 11*, which left Earth in 1973 and reached Saturn six years later. Some gave this pioneering machine less than a 1 per cent chance of success. They were convinced it would be smashed by a collision, either in the asteroid belt or in Saturn's rings. But flight engineers steered the probe right through Saturn's ring plane in September 1979, taking it within 21,000 km (13,000 miles) of the planet's cloud tops without incident. *Pioneer* discovered Saturn's eleventh moon, two new rings, and the fact that Saturn has a magnetic field 1000 times stronger than Earth's. The probe was last heard from on 30 September 1995, just before Earth's motion carried us out of view of *Pioneer's* antenna. It is now somewhere near the outer edge of the solar system, travelling in the direction of the constellation Aquila (the Eagle), and will pass one of the stars in Aquila in about 4 million years.

Voyagers 1 and *2* sped past Saturn in 1980 and 1981, following highly successful encounters with Jupiter 18 months before. Saturn was the last planet that *Voyager 1* visited, and it used the planet's gravity to hurl it out of the plane of the solar system and over Titan's north pole. In November 2003, *Voyager 1* passed another milestone, reaching 90 astronomical units from Earth – 90 times the distance between Earth and the Sun, a total of 13.5 billion km (8.3 billion miles). *Voyager 1* is now the most distant manmade object in the universe – our first interstellar ambassador. It has enough power and fuel to operate until about 2020, by which time it will be almost 22.5 billion km (14 billion miles) from Earth.

Voyager 2 made a close flyby of Saturn, slipping through a gap in the rings and using the planet's gravity to slingshot it onwards to encounters with Uranus and Neptune.

Both Voyagers studied Saturn's rings and moons, but it was *Voyager 1's* close flyby of Titan that mission scientists most eagerly anticipated. The first close-up pictures of the moon were carefully scrutinized for gaps in the clouds that might give a glimpse of the surface – but without luck.

Sixteen years passed before the joint ESA/NASA *Cassini–Huygens* spacecraft was on its way back to Saturn. This six-tonne machine is the size of a school bus, making it the largest planetary probe ever launched by NASA. It also has the most sophisticated 'brain' ever built into a robotic explorer. The brain should enable the probe to sort out problems on its own once in orbit around Saturn, where it will be more than three hours away from mission control's radio signals. In order to pick up enough speed to reach Saturn, *Cassini–Huygens* first looped

around Venus, Earth, Venus again, and then Jupiter. Its complicated journey took seven years, and it reached Saturn in mid 2004.

Cassini–Huygens consists of two probes in one: an orbiter (*Cassini*) and a lander (*Huygens*). The *Huygens* lander will be jettisoned from *Cassini* in December 2004 and will parachute through Titan's methane clouds the following month, relaying panoramic photographs and chemical data to *Cassini*. Two hours after entering Titan's atmosphere, it will hit the surface at 25 km/h (16 mph). *Huygens* is designed to function whether it lands on solid ground or a sea of methane, but its instruments will be able to transmit data for only 30 minutes at most before battery power runs out or the orbiter flies out of communication range. The lander can do most of its work in just four minutes on the surface, hopefully lifting the veil on this

mysterious world.

Cassini will use radar to map Titan's surface and put the landing site in context. It will then spend the next four years studying the rest of the system, making at least 74 orbits of Saturn and observing the planet's atmosphere, rings, and many moons in unprecedented detail.

1. As *Cassini* closes in on Saturn, its view grows sharper and sharper.
2. The two-storey-high robotic pair – *Cassini* (the main craft) and *Huygens* (attached on the right).
3. A true-colour photograph of Saturn snapped by *Voyager 2* in 1981. The moons Rhea and Dione appear as dots to the south and southeast of Saturn, respectively.
4. The orange haze of Titan's thick atmosphere kept the secrets of the surface safe from the gaze of the Voyagers.

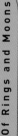

☒ Yvan Grigorev, Flight Engineer
725 days, 8 hours, 22 minutes

I know everyone is concerned about John's health, but I find myself worrying just as much about *Pegasus*. Aerobraking should be easier this time, but the ship took a battering at Jupiter and isn't as strong as it used to be, even after all the repairs. Tom and I have spent the last eight months making a detailed ultrasound survey of critical structural points along the truss, hunting for signs of metal fatigue. Fortunately we found only two weak spots that needed strengthening.

My big anxiety is what might happen when we cross the ring plane, where the dust concentration is highest – space probes have recorded hundreds of impacts a second crossing Saturn's ring plane. The radiation count should be low once we're in the rings, but we'll still be powering up the magnetoshield from the moment we cross Saturn's bow shock. We're already picking up strong radio signals, and the whistling sound reminds me of the approach to Jupiter. It sends a shiver down my spine, and I'll be much happier when we're safely parked in the Cassini gap.

⊞ John Pearson, Flight Medic
746 days, 6 hours, 20 minutes

It's hard to believe we've finally reached this ringed world. There have been times over the last few months when I wondered... Maybe it's the painkillers affecting me, but there is a definite dreamlike quality to this place. I can't imagine a more beautiful planet in the galaxy.

And what a ride to get here! We first crashed into the upper atmosphere five days ago, falling backwards just a few hundred kilometres above the ring plane. I remember looking back towards the weak Sun, which was surrounded by a faint halo of light refracted from ice crystals high in Saturn's atmosphere. For a moment I felt like a Saturnian, watching his own sunrise.

We had to endure only a couple of Gs for 20 minutes or so as we swung around Saturn to the nightside of the planet. Three more gentle dips into the sky over the following days trimmed our orbit

to a point where Tom felt it was safe to attempt an insertion into the Cassini gap. Then, kilometre by kilometre, we nudged *Pegasus* into the ring plane, matching our speed to that of the ring fragments. I watched in awe as the sparkling chunks of rock and ice became clear to the naked eye. First the city-sized pieces, then smaller, house-sized fragments and then those the size of cars and eventually smaller ones – no bigger than people.

Now I find myself transfixed by the view, mesmerized by the endless, slow spinning of the fragments as they glitter like jewels catching the light. The debris seems to extend right down to the cloud tops – the faint furthest ring trailing off right into the top of the planet. Here and there, tiny moons glide through the debris like giant icebergs, casting jet-black shadows across the ring plane's sea of dust.

Nina's going out there today; I'm really jealous. Although we'll be sending back a truckload of images, not one of them could possibly convey what it actually feels like to be here. I feel somehow strengthened by the place, stronger than I've felt for a while. I've stayed in the cupola since we reached the Cassini division. Zoë suggested I move here permanently, and they've all made me very comfortable in my new room with a view.

The rings look almost solid when I look out across them, but staring into the ring I can see just how fragile they are, just a kilometre away. These mighty rings I first saw through telescopes from Earth appear incredibly delicate when you get up close. A temporary spectacle, gradually grinding itself to oblivion. One day it will be gone. I feel very fortunate to be here, alive at this time.

Oh lucky man.

⬚ Zoë Lessard, Mission Scientist
746 days, 8 hours, 10 minutes

For the last two hours I've been helping Nina prepare for her spacewalk – suiting up and running through the hundreds of checkpoints. I've really come to admire her over these months. I thought I had her sussed – I thought she might be the first to crack. But while the rest of us have been struggling with our various problems, Nina has remained extraordinarily focused and stoic. She just gets on with things, and she always finds time to laugh and joke. Now she's about to embark on an incredible exploration – it has to be the ultimate spacewalk and yet she seems so calm and composed. If I were in her shoes I wouldn't be able to contain my excitement, but Nina seems completely unfazed, almost indifferent even, and focused on the tasks at hand.

Once the suit checks were complete, I helped her into the MMU – a kind of jetpack that allows her to move about independently without a tether connecting her to *Pegasus*. It seemed huge, wrapping around her like an armchair. I pulled the controls up in front of her and placed her gloved hand around each propulsion trigger. Then I suited up and went into the airlock with her to keep an eye on things. There was a loud pop and the familiar whooshing noise as the airlock depressurized and the air rushed out. Confident our suits were working, I opened the hatch and watched as Nina floated out. When she was clear of the hatch she span around in the MMU and winked at me – she is just so cool.

🎞 Nina Sulman, Mission Scientist
746 days, 14 hours, 9 minutes

I've just completed my untethered EVA into the rings of Saturn. After being cooped up in *Pegasus* since we left Mars some 28 months ago, it was great to get outside again. And I'd been dying to fly the MMU ever since training – it's a great bit of kit.

My aim was to collect a sample from the B ring – one of the major divisions of Saturn's ring system. As I manoeuvred out of the airlock I felt relaxed and confident, despite the distinctly anxious look on Zoë's face. My confidence didn't last long, however. The moment I took in the view I was hit by a wave of vertigo, something that's never happened to me during EVAs in Earth orbit. Being in the rings gives you a giddying sense of size and

Kirby (left) and Grigorev (right) monitor Sulman's historic walk in the rings of Saturn.

distance, and I felt as if I were floating over some kind of abyss. The immense rings stretched into the distance in front of me and beyond them I could see Saturn's clouds, 112,000 km away. I slowed my breathing and concentrated on my objectives providing a commentary on what I could see.

Although the Cassini division looks empty in photographs, there's actually a lot of dust in it. I couldn't see far into the dense B ring because the dust obscured the view, but I could see where very large fragments stuck out of the plane, casting shadows across the dust. I could also see one of Saturn's small 'shepherd moons' about 80 km away, its heavily cratered, grey surface looking crisp and sharp in the sunlight.

I fired the MMU's left and right jets together and started moving silently towards the dense wall of debris that defines the edge of the B ring.

At first, the ring fragments seemed to be tumbling in a chaotic manner, but after a few minutes I noticed that there was a pattern to their dance – as each pulled and pushed its neighbour, driving their tumbling motion. The fragments bump into each other, sometimes sticking and sometimes breaking up and then reforming. It is impossible to tell if they are constantly wearing away or re-building themselves.

The light came and went as I flew in and out of the fragments' shadows. Despite the flickering sunlight, I could see the surface of the fragments very clearly. Their colour varied with size, but most could be described as translucent grey. My goal was to bring back a chunk of about 50 cm diameter. After a bit of practice with the MMU, I unstowed the sampling tool and flew towards a fragment that looked the right size. The tool worked fine, its pronged ends sticking into the fragment's crusty surface and throwing off a shower of dust.

Now came my second shock. Having secured the sample, I turned around and saw *Pegasus* – it looked tiny, and was about to disappear into the dusty haze. For a second I felt confused and alarmed – had I travelled too far? When you're floating alone in Saturn's rings, it's easy to get spooked. Fortunately, Tom and Zoë worked out what was wrong. My suit had a small electrostatic charge that was attracting particles of dust – I was slowly getting covered in dirt. I wiped the helmet and the visibility improved dramatically. I wasn't sure how the dust might affect the MMU, so I decided to return to *Pegasus* as soon as possible.

The EVA had been a success. I had 'walked' in the rings of Saturn

Sulman amongst the ring fragments, 112,000 km from the cloud tops of Saturn.

and, brought back a sample. It was hard not to feel elated at the triumph of the day. I flew back to the ship fully expecting a big celebration. How wrong I was.

⊠ Tom Kirby, Commander
746 days, 18 hours, 55 minutes

Flight Medic John Pearson passed away at 11.15 a.m. on the 746th day of the mission. We all knew it was coming. We'd lived with it for months, but it still came as a shock.

I've known John for as long as I can remember and we've been through so much together. A big part of my life has been taken away with John's passing, and I know the rest of the crew feel the same.

John was hugely proud to have flown on this mission and especially pleased to reach Saturn – a childhood dream. We discussed it and decided to give him a burial I know he would have wanted. I suited up and floated his body out into the Cassini division. In a very real sense he has become part of Saturn.

We now face a choice: do we continue to Pluto or turn back to Earth? I'm giving everyone time to think about it. Whatever we do, the decision must be unanimous.

⊟ Zoë Lessard, Mission Scientist
749 days, 15 hours, 51 minutes

As usual, Control don't seem to grasp how hard the last few days have been for us. We knew it was coming, but it's impossible to prepare yourself. The loss I feel is impossible to put into words. John was as much a part of my life as anyone could be, and his passing leaves me feeling empty. The slightest thing can set me off and I have to leave the room. On *Pegasus* each of us has colour-coded belongings; all John's things are purple, and even the sight of this colour is painful. I keep looking back to the spot where Tom placed his body. I don't know what I expect to see. I guess John would have been thrilled to know he was going to be buried here in the rings. What better mausoleum could anyone hope for – visible for all on Earth to see.

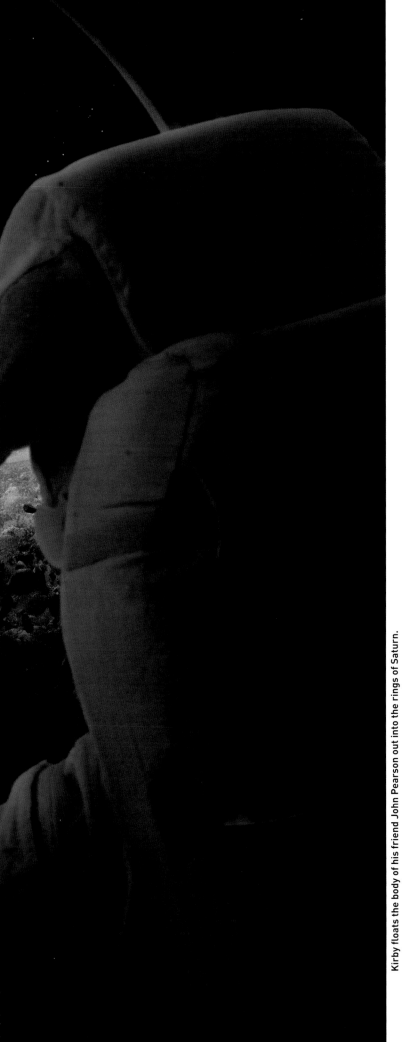

Kirby floats the body of his friend John Pearson out into the rings of Saturn.

Tom is the only person I can talk to about it. He feels the same way as me, and we've been consoling each other. Yvan isn't a talker and prefers to just get on with his work. Nina too – she turns the conversation around to her experiments whenever we chat. But I suppose we've all got to get through this in our own way.

There are big decisions to make. I'm inclined to abandon the second phase of the mission and return to Earth. We've done more than enough to justify our flight, and it's a wonderful legacy to return to Earth with. Tom is undecided, and I don't know how the others feel.

We've asked for time to think it over. It has to be a joint decision, but it can't wait much longer – if we're going to attempt an Earth return trajectory, we need to initiate the flight procedures to make it happen soon.

Nina Sulman, Mission Scientist
751 days, 8 hours, 18 minutes

The sample I retrieved from the B ring has been floating in the cryogenic vacuum chamber for five days now. Its surface is grey and glassy, corroded by sunlight. I should be able to work out its age from the amount of solar erosion, and I'm guessing it will turn out to be about 120 million years old, but it could be much older if it's spent part of its life buried in rubble on one of Saturn's smaller moons. The origin and evolution of Saturn's rings is written into this sample's atoms, and I intend to play my part in teasing out every clue it contains.

John's death has made me all the more determined to get the most out of this mission and press on to Pluto. What discoveries would be lost if we gave up now and returned to Earth? I doubt we'd build another *Pegasus* for a long time, and even if we did, would it be so lucky as to get this far? With every fibre of my being, I believe we should continue the grand tour. John would never have wanted it abandoned – we owe it to him to go on.

☒ **Yvan Grigorev, Flight Engineer**
752 days, 14 hours, 15 minutes

For the first time since leaving Earth, I'd say we have a really good chance of reaching Pluto. *Pegasus* has proved herself. She's come through the worst the solar system can throw at her – a flight through the solar corona, an aerobrake through Jupiter's atmosphere, and a month inside its dreadful magnetosphere. I don't think there's anything the outer solar system could throw at us that would be more serious.

It's ironic that *Pegasus* can do it but the humans that built her aren't up to the task. I never thought any of us would perish in the way John did, wasting slowly away. I'd always thought that any deaths would be sudden and violent, caused by a crash or an explosion. But we're such fragile souls, unsuited to life outside Earth's protective atmosphere, and much harder to repair than *Pegasus*.

Fragile as we are, we have a unique opportunity to reach farther out into space. We signed up and trained to go the full distance, and John's death shouldn't change anything. Some tribute it would be to him if we turned back now. Where would we be if such setbacks had stopped our species from exploring Earth, crossing the oceans, taking to the air or reaching for the stars? We would be betraying humanity if we turned back.

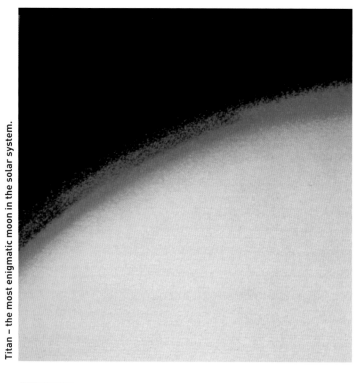

Titan – the most enigmatic moon in the solar system.

✴ **Tom Kirby, Commander**
760 days, 15 hours, 50 minutes

Our work around Saturn is coming to an end. If we go on, Titan – Saturn's largest moon – is our final objective here. It is a world I would love to unveil. Like Venus, Titan is shrouded in thick cloud that hides its surface. The fact that Titan has an atmosphere at all is intriguing – a world this size should not have enough gravity to hold on to gases for long, so something must be replenishing Titan's atmosphere. There's no sign of volcanic activity, and although the surface is –178 °C, it isn't frozen solid. Titan is a world of hydrocarbons. Giant raindrops of gasoline probably fall from its overcast skies, and rivers of the stuff might wind their way towards rolling hydrocarbon oceans. It's the only world other than Earth that might harbour such oceanic vistas. We only plan to drop a robotic probe down there, but what a triumph it would be to capture a glimpse of Titan's surface.

Somebody once said that the greatest risk is not to take one. To return to Earth now would be to give up, to accept failure and to disappoint millions of people. We've risked our lives to come this far, and I'm prepared to risk mine to continue. But I don't think our decision is going to be unanimous.

Sulman in awe at the sight before her as she drifts silently in Saturn's rings.

Pegasus makes its rendezvous with a food/fuel dump in orbit around Titan.

When Moons Collide

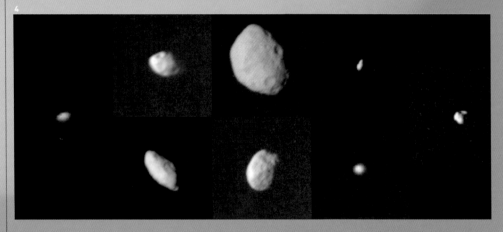

suggests this may have happened as recently as 100 million years ago. At first the debris would have been scattered in a cloud, but this was gradually tugged by Saturn's gravity into a thin plane around the equator. Further influenced by the gravity of Saturn's moons, the debris has been 'shepherded' into seven discrete rings, which are lettered in order of their discovery. Gian Domenico Cassini discovered the A and B rings in 1675. The fainter C and D rings were identified later by telescope, and the even fainter E, F and G rings were discovered by *Pioneer 11* and the Voyager probes.

The Voyagers also discovered curious details in the rings that are not visible from Earth, such as twists in the F ring caused by gravitational tugs from large, asteroid-sized fragments in the rings. These are Saturn's 'shepherd moons', named after characters in Greek mythology: Atlas, Prometheus and Pandora. Recent research suggests these tiny moons soak up and release debris into the ring plane, slowing down the breakdown of the rings into dust. If this theory is correct, the shepherd moons are really just orbiting piles of rubble, and their presence could make Saturn's rings last far longer than was once thought. The capture and recycling of ring debris is something that the *Cassini* spacecraft will study during its visit.

Saturn's ring system is one of the wonders of the solar system. The rings are four times wider than the planet itself, extending from about 7000 km (4300 miles) above the atmosphere to 420,000 km (260,000 miles) out, which makes the outer rings further from Saturn than the Moon is from Earth. Yet the rings are barely more than a kilometre thick – equivalent in proportion to a sheet of tissue paper stretched across a football pitch. The total amount of material in the rings is surprisingly small. Massed together, it would form a ball less than 100 km (62 miles) wide.

The debris in the rings is bright and probably consists mostly of water ice, though there is also a lot of dust and rocky particles coated in ice. Fragments range in size from tiny grains to blocks of ice as big as cities.

The rings are thought to have formed during a massive impact, perhaps between a comet and a moon or between two moons. The fresh appearance of the ice

1. Gian Domenico Cassini's sketches of the rings made in 1676.
2. The far side of Saturn, taken by *Voyager 1*.
3. A highly enhanced colour view of the rings reveals subtle chemical differences.
4. A clutch of smaller satellites of Saturn as viewed by *Voyager 2*.

Chapter 5

Mysterious Bodies

Saturn has at least 30 moons in its gravitational clutches. The nearest is just 133,600 km (83,000 miles) from the planet; the farthest is more than 13 million km (8 million miles) away. Many are small, irregular hunks of rock and ice, just a few tens of kilometres across and thought to be captured asteroids. Epimetheus and Janus, 151,400 km (94,100 miles) from Saturn, are particularly interesting because they appear to be two parts of a single moon that broke apart.

Farther out are a number of better-mapped moons. Mimas is about the size of

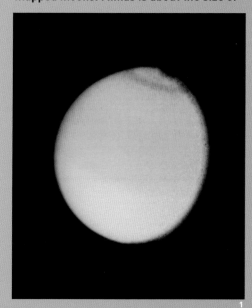

England and most famous for it's resemblance to the Death Star in *Star Wars*, with a large crater called Herschel dominating one face. The crater is so vast that it's a wonder the impact didn't smash the moon apart.

Enceladus is half the size of Mimas and couldn't be more different. It has few craters and a curiously fresh, smooth surface. It is the most reflective moon in the solar system, bouncing back almost 100 per cent of the sunlight that hits it. Farther out are the bigger moons Tethys, Dione, Rhea and Iapetus – all cratered worlds of ice, with grooved surfaces. Iapetus has the strangest surface of all, with one hemisphere bright and icy and the other coated with an unidentified black substance, the depth, chemistry and origin of which are a mystery; the *Cassini* probe will make a close flyby to investigate.

Saturn's biggest moon, Titan, is about the size of the planet Mercury. It's big enough to be seen with a small telescope from Earth, and it orbits Saturn every 16 days or so at a distance of 1.3 million km (800,000 miles). The first hint that Titan had an atmosphere came at the start of the twentieth century, but it wasn't until the Voyager flybys that scientists realized how dense Titan's atmosphere is. Atmospheric pressure on this moon is about 60 per cent greater than on Earth – roughly the same pressure you'd find at the bottom of a swimming pool. The air is so thick and the

gravity so low (about a sixth of Earth's gravity) that you could probably fly on Titan if you strapped cardboard wings to your arms. Titan's air is mostly nitrogen, and UV light seems to have turned the upper atmosphere into an impenetrable smog. The surface temperature is about −178 °C (−288 °F) – and that's about all we know for sure about what it's like below the clouds.

Infrared images from Hubble and radar maps made from Earth suggest Titan has a varied terrain, with higher and lower regions. There are also large, bright features, including one the size of Australia. This might be a continent made of water ice, solid carbon dioxide, rock and tar, surrounded by an ocean of liquid methane or ethane. Purely by coincidence, the *Huygens* lander will drop close to the edge of one of these bright regions.

Some scientists think Titan is similar to the primordial Earth. Though the moon is too cold for liquid water or for complex life, its tar-caked surface might hold clues to the origin of carbon-based life on Earth.

1. A *Voyager 2* image of Titan shows very little detail in the clouds. The extended haze, composed of very tiny particles, is seen around the satellite's limb.
2. The heavily cratered moon Mimas, taken by *Voyager 1*. The smallest features that can be seen are about five miles across.
3. Many impact craters – the record of the collision of cosmic debris – are shown in this *Voyager 1* mosaic of Saturn's moon Dione. The largest crater is less than 100 km in diameter and shows a well-developed central peak.

To the Edge

PLUTO

- DIAMETER: **2324 KM (1444 MILES)** ⬚ MASS (EARTH = 1): **0.002** ⬚ GRAVITY (EARTH = 1): **0.07**
- AVERAGE SURFACE TEMPERATURE: **−233 ºC (−382 ºF)** ⬚ DAYLENGTH (ROTATION PERIOD): **6.39 EARTH-DAYS**
- YEAR: **248 EARTH-YEARS** ⬚ AVERAGE DISTANCE FROM SUN: **5.9 BILLION KM (3.7 BILLION MILES)** ⬚ RINGS: **0** ⬚ MOONS: **1** ⬚

After the death of John Pearson we expected the crew of *Pegasus* to turn back. They had lost their doctor, their counsellor and their friend; time to call it a day, we thought. But to our surprise, they decided to carry on. We didn't know if the decision was based on pure determination to finish the mission, respect for John, or just madness, but from the public's point of view, the crew were heroes. From that point on there was no more bad press, and though interest in the project came and went over the next three years, there was nothing but goodwill for those four brave souls who were travelling farther and farther from home.

With their decision came a list of proposals, or 'demands', as one colleague called them. These involved some fundamental changes to protocols and research goals, not least the complete reconfiguration of the telescope they were to set up on Pluto. The science team grumbled, but in retrospect the crew's ideas were spot on. I detected Tom's hand in every line, gently guiding and managing his wounded team. When we devised the itinerary for the grand tour we had addressed every conceivable issue and planned the mission in minute detail. This left little for the crew to do but tinker in the lab. Understandably, they wanted to do more – they wanted to control the mission. Perhaps they felt they'd earned the right.

We held a memorial service for John at Control and transmitted it to *Pegasus*. The crew also conducted their own private celebration of John's life. We suggested they take some time off, but they seemed eager to get back to work – perhaps keeping busy was the best way to deal with things. The next task was to drop a probe into the atmosphere of Titan, one of Saturn's moons. Unfortunately the probe malfunctioned and disappeared into the orange clouds without trace. The crew weren't noticeably upset, however, and got on with the task of radar-mapping Titan's hidden oceans and continents. They saw the probe's failure as a vindication of the decision not to send humans there.

Then came the last, longest leg of the mission – the journey to the outer solar system and Pluto. It would take at least 600 days and called for a slingshot from Saturn to lift *Pegasus* slightly up and out of the plane of the solar system. The ship continued to function superbly, but the nature of the challenge facing the crew had now changed. Crossing the empty wastes of deep space is as much a psychological challenge as a technological one – would these four souls hold it together long enough to reach the very edge?

Alex Lloyd
Chief Scientist, *Pegasus* Mission

 Zoë Lessard, Mission Scientist
Mission elapsed time: 780 days, 18 hours, 45 minutes

- -

Watching John's memorial service transmitted from Earth a few days ago, it felt like an archive from another generation. The 90-minute delay between us and Earth might as well have been 90 years. Our lives on *Pegasus* feel so detached from events back at home.

At the end of the prayers, bells were rung and three fighter

Pegasus performs a Trans-Pluto Injection Burn – to propel the mission away from Saturn and on towards the outer solar system.

aircraft flew past in formation. We sat in silence for 15 minutes. Then we moved on. We had work to do.

The only way to think about the voyage to Pluto is as a return trip. It's not just a 20-month flight to Pluto – it's a 40-month round trip, with a short bout of activity in the middle when we reach the planet. Although we've already spent more than two years in space, the next three years seem to stretch out like some interminable sentence. We've never been more than a few months away from our next encounter, but now the months of empty

space stretch out ahead of us. Even with schedules to swot up on, simulations to run, the task of redesigning the Pluto telescope, and the usual maintenance chores, there will be time off.

I've noticed that our sense of priorities has changed. We've been playing zero-G tennis on *Pegasus* since the voyage began, but recently this pastime has become as important as maintenance chores. The sport was invented on shuttle flights in the 1990s, and I guess it's a combination of gymnastics and tennis. The ball is a lump of gaffer tape, and we use clipboards as rackets.

There's no net, but you gain points if your opponent misses a shot. After that it's a free-for-all – forehand, backhand, upside down, overhead, off the walls, whatever. Yvan has lead our 'Grand Tour Grand Slam' tournament since we started out – but there's still a long way to go and a lot more games to play.

✴ Tom Kirby, Commander
800 days, 12 hours, 30 minutes

Pluto is a tiny target – only two-thirds the size of the Moon, and not much more than twice the size of the largest-known asteroid, Ceres. When we left Saturn it was a staggering 4.2 billion km away. Our best telescope could make out only a tiny, indistinct disc with vague light and dark patches. It's no easy matter to rendezvous with this tiny world. And what makes Pluto even harder to reach is its peculiar orbit, which takes it out of the 'ecliptic' – the plane in which all the other planets orbit.

The gravity slingshot from Saturn, combined with a lengthy engine burn, has flung *Pegasus* onto a course for Pluto that is now climbing above the ecliptic. There's something different about this trajectory – it feels as though we are leaving the solar system behind and heading into the deep unknown.

This region of the outer solar system is not fully charted, but we do know the main landmarks. There's a minor 'planet' between Saturn and Uranus, called Chiron. It's four times smaller than Pluto, a mere 970 km wide, and takes 66 years to orbit the Sun. Farther out, of course, are the ice giants' orbits Uranus and Neptune and another asteroid belt, and beyond them the Kuiper

Kirby and Sulman run through simulations for Pluto.

Lessard floats in her sleep module during a routine maintenance period when the centrifuge is powered down.

belt – a vast expanse of dark, icy fragments left over from the formation of the planets.

But it's not a collision with a lump of ice that worries me. There are more subtle dangers on this part of the mission – like boredom, depression and cabin fever. We've now been in space for longer than anyone else, and by the time we get back from Pluto we'll have more than tripled the existing career record of 747 days, accomplished by cosmonaut Sergei Vasiliyevich Avdeyev.

No-one knows what being cooped up for so long will do to us – especially out here, where we could start to feel lost or abandoned. With John gone, it's up to me to keep an eye on the crew's wellbeing and state of mind. For guidance and inspiration, I'm re-reading Ernest Shackleton's diaries, which he wrote during his fraught Antarctic expedition in 1915. I wonder what he would have made of this leadership challenge.

🙾 Zoë Lessard, Mission Scientist
840 days, 22 hours, 20 minutes

Today it occurred to me that I haven't been outside for almost two and a half years. I mean properly outside. Of course I got to land on Venus and walk on Io, but in a bulky spacesuit you don't really feel like you're outside. You see the world through a window and touch the ground through thick gloves or boots. The feel of wind in my hair and rain on my face and the smell of fresh blossom are just memories. And I miss them.

There's a constant stream of TV channels from Earth, should we want them, but it's no substitute for being there. I prefer to immerse myself in e-mails, and I seem to spend more time communicating with my fingers these days than with my vocal

Sulman continues with her agricultural programme – perfecting crops to supplement their diets.

cords. Sometimes I forget that I'm not using my voice for real – I can hear it so clearly in my head as I type, engrossed in my distant, delayed conversation. It takes at least four hours to get a response from Earth, and each day adds another 23 seconds to the delay. It might sound like a big disruption, but with multiple conversations on the go, you don't notice the delay so much. I can easily forget where I am, transported in my mind back to Earth – to the homes of friends and family wallowing in the minutiae of their lives. It's a welcome distraction from my own life, trapped in here with nowhere to go.

🔲 **Nina Sulman, Mission Scientist**
870 days, 14 hours, 50 minutes

For the first time on the flight there's a chance to stop what we're programmed to do and just sit back and ponder what an amazing adventure this is. When everyone's asleep I sometimes float into

the cupola to look at the cosmos. Out of the glare of the shrinking Sun, our galaxy – the Milky Way is clearly visible arching overhead – a brilliant, milky wash of a hundred billion suns. It's a view I never get tired of.

A few days ago Yvan brought his saxophone into the cupola, and his rendition of 'Fly Me to the Moon' got us all singing. I wondered what we must have looked like from the outside of *Pegasus*, all floating under a small glass dome singing swing in the outer reaches of the solar system.

🔲 **Nina Sulman, Mission Scientist**
953 days, 12 hours, 35 minutes

This part of the solar system is pretty empty, but there is still plenty to explore. Zoë is hunting for new moons of Pluto; astronomers are convinced they exist. Any such moons would probably be small, dark bodies from the Kuiper belt. The only way to find them is to take multiple photographs of the sky around Pluto and then look for tiny differences caused by the movement of any moons across the backdrop of fixed stars. Pluto itself was discovered by a similar technique back in the late 1920s.

I've taken over John's studies of the effect of cosmic radiation and low gravity on our bodies. I continue to monitor everybody's bone and muscle strength regularly, checking most importantly for loss of bone density (osteoporosis). We maintain our bone health using a machine that vibrates our limbs for a few minutes each day, and the treatment seems to be working fine.

Tom and Yvan are keeping up a rigorous daily exercise regime to make sure they're strong enough for the construction job they'll have to perform on Pluto. They both devote a lot of time to the redesign and testing of the telescope. We hope it will be up to the task of imaging other Earth-like planets around distant stars from the surface of Pluto. Yvan has cannibalized components from spare parts around *Pegasus* for the conversion. Even with all the time they've got, rebuilding the telescope is a daunting task. When they're not doing that, they are practising their Pluto landing in the simulator. Zoë keeps trying to kill them. They've both died four times so far – but it's for their own good! They'll thank her when they're coming in to land for real in a year's time.

I'm also continuing with my own agricultural programmes, trying to improve the lettuces, tomatoes, spring onions and radishes that I grow in soil-free 'hydroponics' trays dotted around the ship. These fresh foods supplement the dried supplies we picked up at Saturn. Tom is a keen cook – he uses the bread maker to blend soya beans, tofu and other ingredients into experimental foods. It's not always very tasty, but it's a welcome change from the usual fare.

⊞ Tom Kirby, Commander
1211 days, 21 hours, 8 minutes

Low gravity can have some strange effects on the human body. We've all noticed that food tastes better in the gravity modules than in zero G. Nina says it's because weightlessness allows fluid to build up in our heads, giving the sensation of a head cold. Food that's tasty in gravity can be completely bland in weightless parts of the ship, so we use spicy sauces to liven it up.

Earlier in the mission, when the gravity simulator had to be switched off for several weeks, I saw Nina adding Tabasco sauce to everything she put in her mouth, and even now she seems to consume an extraordinary amount of the stuff. At the rate she's getting through it, our supply will run out before the end of the mission. I've decided to confiscate the last reserves and ration it over the coming months, restricting it to periods when we don't spend sufficient time in the gravity simulator to restore our sense of taste. Let's hope we make it to Pluto before the Tabasco runs out completely.

⊞ Tom Kirby, Commander
1350 days, 23 hours, 5 minutes

Everyone's getting excited about Pluto and Charon – it's hard to believe we are so close. My fears about how well the crew would cope with the long journey have so far proved unfounded. Although everyone has found their own corner of *Pegasus* where they can spend time alone, nobody has become too withdrawn or retreated from social gatherings. Yvan is perhaps the most distant, but I know he maintains a furious amount of communication with friends, colleagues and family back home. Boredom and loneliness have always been there, but no-one has lost their sense of purpose on this mission.

Kirby and Grigorev in polar orbit around the Pluto–Charon system.

The Search for Planet X

During the nineteenth century, astronomers had noticed that the orbits of the outer planets Uranus and Neptune were disturbed. The exciting interpretation of these wobbles was that there was a huge ninth planet that people had somehow failed to spot. The search for the mystery planet was on.

The hunt was led by two US astronomers: Percival Lowell and William Pickering. Lowell began searching in 1905 for 'planet X', which he reckoned was seven times Earth's mass and took 282 years to orbit the Sun. Pickering joined the hunt in 1909. He thought 'planet O' was no more than twice Earth's mass and took 373.5 years to orbit the Sun.

Lowell and Pickering took more than 1000 photographic plates between them and spent years scrutinizing them, but to no avail. Lowell was a wealthy businessman and had built an observatory in Flagstaff, Arizona, to indulge his hobby of observing the planets. He died in 1916 before a ninth planet was discovered, but he left money in

his will to continue the search. Backed by extra money from Lowell's brother, a third (and, some thought, pointless) search began at Flagstaff in 1929. The observatory's director appointed a young amateur astronomer called Clyde Tombaugh to do the work.

Tombaugh's task was to take pairs of photographs of tiny portions of the sky on different nights, and then compare each pair using a special viewing device that allowed him to flick between the two images. Anything that had moved against the unchanging background of stars – like a planet or an asteroid – would quickly

become apparent. Computers use the same technique today to look for asteroids, but in Tombaugh's time it had to be done entirely by eye. It was a very tedious task and there were no shortcuts.

Day after day, he sat in front of his viewing device studying photograph after photograph. Each photographic plate might

contain anything from 50,000 to a million stars and took up to a week to examine properly. Tombaugh found it exhausting and could concentrate for only 6 hours a day. The photographic plates piled up, but there was no way he could rush.

Occasional flaws in the plates caused heart-stopping false alarms, and almost every plate contained a handful of asteroids. Tombaugh knew that planet X was likely to move only a few millimetres between plates, rather than the centimetres the asteroids moved, but they still distracted him.

Tombaugh's amazing patience eventually paid off. In February 1930, after nearly a

year of searching, he discovered a speck of light that moved only 3–4 mm. 'A terrific thrill came over me,' he later recalled. 'I switched the shutter back and forth, studying the images.' He was so sure it was planet X that he checked his watch and made a note of the time. It was 4.00 p.m.; 'Planet suspect', he noted in his log book.

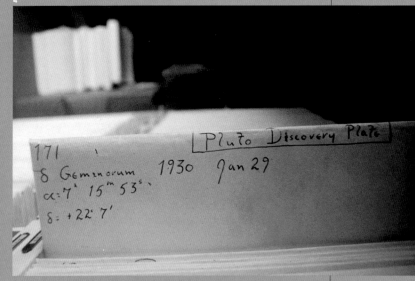

No further examination was done for a considerable...
used the comparator for measurement of positions on the...
identification of Pluto images on plates that he made...
following nights for several months.
Examination of this pair of plates was resumed a...
devote only a small part of each day, and examination...
was not completed until June 9, 1930.
The Ottawa object was watched for on the remain...
nothing else of a transneptunian nature was foun...
...es of Examination :—
8 asteroids "A".
7 variable stars "V" (+ 3 or 4 others not marked on this pl...
12 temporary objects (on one plate only) which may be...
...different suspicious objects (transneptunian) of 17 t magd. were...
...ll cases, from checks on other plates — on one plate a pair...
No comets.
Planet "X" (Pluto) at last found !!!

171
8 Gemenorum   1930   Jan 29
α = 7ʰ 15ᵐ 53ˢ
δ = +22° 7'

Pluto Discovery Plate

After confirming the results with other plates, Tombaugh told his boss Vesto Slipher, who wisely advised Tombaugh to keep quiet about the discovery until more checks had been made. The tiny speck was fainter than predicted – about 250 times fainter than Neptune – but Slipher was sure it was planet X. On 13 March 1930, to coincide with Percival Lowell's birthday, the news was released, and headlines around the world proclaimed 'ninth planet found'.

1. Percival Lowell (1855–1916), passionate about Mars but equally famous for his prediction of the existence of a ninth planet outside the orbit of Neptune.

2. The Pluto discovery plates taken by Clyde Tombaugh six days apart. When they were compared planet 'X' was seen to move between them (see arrows).

3. Tombaugh's notes on the sleeve of the plate – noting that 'Planet X' was 'at last found'.

4. A Pluto discovery plate in store at the Lowell observatory in Flagstaff, Arizona.

5. Clyde Tombaugh at the time of the search enters the telescope dome, holding a wooden photographic plate under his arm.

◉ Zoë Lessard, Mission Scientist
1385 days, 18 hours, 10 minutes

My God, it's pink! The surface of Pluto from orbit looks pink! It's a subtle pink, but it's definitely there. Our best guess is that the colour comes from the breakdown of some kind of carbon-based substance in the surface ice.

We reached orbit 14 days ago, three years and nine months after leaving Earth. It was a joy to be so close to a planet again – we had felt like sailors who'd been away at sea and had finally made landfall. In close consultation with Control, we spent the first few days charting the surface and looking for a landing site.

Tom and Yvan left for the surface this morning and made a safe landing, thanks no doubt to their many trial runs in the simulator. Now there's just me and Nina on board, and it feels strange having the whole ship to ourselves.

Pluto isn't at all what I'd expected. Its surface is extremely varied, with crisscrossing lines where pressure has broken the ice into ridges and walls. Some areas look rocky, and there are black patches with a glassy sheen – perhaps coated in a fresh frost of frozen nitrogen. For the past couple of decades, Pluto has been rushing away from the Sun on its oval orbit, and the planet has been getting colder. The surface is chilling rapidly, and the tenuous atmosphere is freezing to the ground as a thin frost. This planet must have the most extreme seasons in the solar system.

We've been able to see Pluto and its moon, Charon, from the command centre on *Pegasus* for about six months. At first they looked like a pair of stars, but about two months ago they started to appear as spheres – half in light, half in darkness. From then on they became clearer every day. Charon is more than half Pluto's width, making it the largest moon relative to its parent planet in the solar system. The pair swing around each other every six days or so.

We had crossed the void between Saturn and Pluto in a curving flight path that chased after Pluto and Charon in their orbit around the Sun. For most of the trip we were travelling 10 times faster than Pluto, so we had to lose a lot of speed in order to enter orbit. Pluto doesn't have enough atmosphere for an aerobraking manoeuvre (its air density is a 30,000th of Earth's), so we turned *Pegasus* around and fired the engine in reverse to bring down our speed until it matched Pluto's.

Pegasus in a polar orbit around Pluto. Charon is in the distance.

Pluto takes 248 Earth-years to orbit the Sun. Its axis is tipped right over, so the planet rolls along like a barrel rather than spinning like a top. Its north and south poles don't point north or south at all. Instead, they take turns facing the Sun or facing away for decades at a time. We've arrived at the point in Pluto's orbit when the north pole is on its way from day to night. The landing site is in daylight for about five Earth-days at a time, followed by about 30 hours of total darkness.

In John's absence I gave Yvan a crash course in geology during this last leg of the voyage. He's a fast learner, and it was fun to conduct the preliminary mapping survey of Pluto together when we reached orbit. Part of the fun of mapping somewhere for the first time is that you get to suggest names for various places. We've already used our own names: Kirby Crater, Lessard Canyon, Mount Nina, and Grigorev Geyser Plains. We've also been through our families and a bunch of famous astronomers and scientists who devoted their lives to searching for or studying this planet. The landing site we picked is on Lowell Plateau, which is an elevated plain about 50 degrees north of the equator. It looked smooth and stable, lacking the rugged ice ridges of the Tombaugh Plains.

Tom and Yvan undocked on the nightside of Pluto but quickly crossed to the dayside, allowing me to monitor their landing. *Pegasus* orbits Pluto once every 40 minutes, and we're in communication with the lander for just over half that time. It would have been less, but Yvan rigged up an ingenious system that bounces radio signals off Charon, extending our contact by a few minutes. Tom and Yvan have enough supplies to spend ten days on the surface – we hope that's enough time to get the telescope up and running.

Despite my doubts about this part of the mission, I'm glad we came. This frozen, remote world is a tranquil and calming place to be in orbit around.

☒ Yvan Grigorev, Flight Engineer
1381 days, 20 hours, 4 minutes

Tom and I touched down on Pluto this morning, and it's a delight to be back on solid ground again, even if the gravity here is a fraction of that on Earth.

Our lander *Clyde* is a spiky, five-legged flying machine with a single rocket for both landing and returning to *Pegasus* – a simple design that means less can go wrong.

Undocking from *Pegasus* in the dark was nerve-racking – we had to rely on instruments alone to exit the cargo bay safely. Once we were clear of the ship, we fired up the engine and began our descent. Still over the nightside of Pluto, we couldn't see much through *Clyde*'s small windows, though the surface was faintly illuminated by moonlight from Charon.

Flying backwards, we fired the engine again to slow our orbital speed and shed height. The Sun rose behind us, and now we could see long shadows on the surface, cast by mountains and pressure ridges. I started calling the altitude readings as we dropped through the 10-kilometre mark, suddenly aware of how rapidly we were falling.

We were dropping at more than 30 m/s, and within five minutes we'd fallen another 8 km. The humps and ridges of Pluto's rugged surface flew past the window, and Lowell Plateau loomed into view earlier than we'd predicted. We were in danger of overshooting the landing site, so Tom took manual control and fired the engine again to check our westward drift. I felt my harness strain with the sudden force of the deceleration. We were back on course, but now we were falling even faster, and *Clyde* lurched again as Tom turned the lander and reduced our descent speed. I continued my running commentary of our descent.

From 100 m up I could see the surface very clearly. The flat ground appeared to be a dense shelf of ice, with dark, rocky outcrops protruding through. At 10 m, Tom slowed our fall to a hover and peered through the window as he tried to bring all five footpads down onto firm ice. We still had plenty of fuel, so he could afford to be choosy. Out of my window I could see *Clyde*'s shadow cast across the ice in the morning Sun. Beneath us the ice glowed orange in the light of the engine, and nitrogen frost began vaporizing into swirling mists that snowed back to the ground a few metres away. It was probably the highest recorded temperature on Pluto for two or three billion years.

The green contact light flickered on as we settled very gently on the surface. Very calmly, Tom turned off the engine and ran through the safety checklist. I looked out across the ice fields towards the shrunken dawn Sun. We were on the surface of Pluto.

⊞ Tom Kirby, Commander
1381 days, 21 hours, 20 minutes

Control didn't know we'd landed safely for five hours, and their congratulations took another five hours to get back to us. So, for ten hours, there were only four of us to share in this triumphant touchdown on the solar system's outermost world. We soon lost Zoë's voice as *Pegasus* slipped over the horizon and the static hiss of radio silence filled our headsets. It was just Yvan and me alone on the icy wastes of Pluto.

Despite some slight navigation errors, the landing went well. We weren't too exhausted by the ordeal and decided there was no need for a rest period before our first EVA. Suiting up took about two hours, as each of us ran through the other's checklists. Yvan gave me a thumbs up, and I started to depressurize the cabin. Pumps whirred into action, inhaling the air and storing it for our return. A green light near the hatch confirmed the pressure had equalized with the near-vacuum outside, and I broke the seal on the door. With a short whoosh it popped outwards and swung up out of sight.

A cool, subdued light flooded into the cabin. I turned around and began to back out of the hatch, with Yvan guiding me as I felt for the ladder with my boots. All I could see was Yvan's feet and the floor as I made my way down, hunting blindly for each rung. I dropped confidently from the last rung onto the footpad, enjoying the ¹⁄₁₅th gravity that pulled me gently to the ground. I turned around to face the view: a vast, icy plain shined brightly under a black sky and stretched towards the horizon in every direction. Long shadows accentuated every wrinkle and ripple in the ice. The Sun looked small, but it was dazzlingly bright nonetheless. I stood still for a few seconds, mesmerized by the serenity and beauty of the place. Then Yvan broke the moment with a loud hoot of laughter as he started to bounce around enjoying the low gravity.

⊞ Tom Kirby, Commander
1385 days, 18 hours, 50 minutes

We've been here four days now, and I've grown particularly fond of Pluto's big moon, Charon, which is a constant feature above the western horizon. It must be at least eight times wider than a full Moon on Earth, and it stays in the same part of the sky all the time because of the way Pluto and Charon are locked in orbit holding

Kirby (foreground) and Grigorev survey Pluto's icy plains. The distant Sun over four billion miles from here is not much brighter than a star.

the same face towards each other. Although Charon doesn't move, it slowly waxes and wanes as it moves across the sky relative to the Sun. During EVAs I find myself searching for Charon to get my bearings. There's no magnetic field here, but Charon serves as a perfect compass point, marking west all the time.

We've been working long shifts to make the most of the remaining daylight. When we landed we had about 120 hours of daylight ahead of us, but now we've only 15 hours left. After that we face the 30-hour autumnal night – not ideal for exploration, but we have no choice. Fortunately, Charon will cast some light over the landing site, and *Clyde* has floodlights.

The telescope build is going well, and moving the bulky components into place is usually easy. I sometimes feel like superman, taking giant leaps while carrying huge mirrors in one hand. The suits do a great job keeping us warm. It's –228 °C here, and that will fall to –238 °C at night, which is cold enough for most gases to become totally solid. Our boots are particularly impressive – I can spend all day standing on ice and my feet never

get cold. But sometimes I stumble and my hands land on the ice – and then I can really feel the cold through the palms of my gloves. The gloves, like the rest of the suit, are inflated to a high pressure, which forces your fingers apart. Whenever you hold something you're squeezing a pressurized glove, and after hours your hands get very tired.

We haul the components from site to site on specially designed sledges. The runners are heated with electric elements to melt the nitrogen ice helping them slide smoothly, but they don't always work. Sometimes they lose heat, and the sledge freezes to the spot without warning like a stubborn mule. The abrupt halt can easily unbalance you in the low gravity and send you tumbling to the ground. In the first day or two we fell over a lot, and I was worried we'd damage our life-support packs. Luckily we hit the ground so slowly that the solid ice is quite forgiving.

Clyde is our new home, a warm and cosy refuge held clear of the frozen surface by its gangly legs. We sleep well in our hammocks, exhausted by the day's activities. Contact with *Pegasus* comes and

To the Edge

153

A Black-and-white World

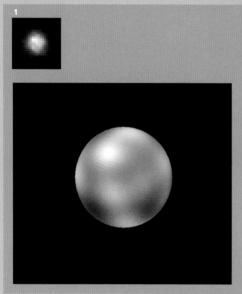

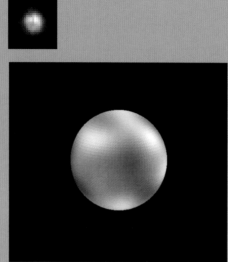

The discovery of the ninth planet was a huge news story in 1930, but the public were more interested in naming the planet than finding out what it was like. Venetia Burney, an 11-year-old girl from Oxford, came up with the winning suggestion: Pluto, Greek god of the underworld, who had the appropriate quality of being able to render himself invisible. Venetia Burney's grandfather, a librarian at Oxford's Bodleian Library, telexed the idea to the International Astronomical Union. Vesto Slipher, the director of Lowell Observatory, liked it – not least because the abbreviation PL also happened to be Percival Lowell's initials.

There were two other names on the shortlist: Minerva, the Roman goddess of wisdom; and Cronus, the son of Uranus and father of Neptune. But the name Minerva was already taken by an asteroid, and Cronus was suggested by a rather unpopular astronomer. So Pluto it was, and Venetia was credited with naming the

planet. Disney's dog appeared in cartoons later in the year, but Venetia was not credited with his name.

Precious little was known about Pluto at the time of its discovery. Even its size was a puzzle – it seemed much too small to make Uranus and Neptune wobble in their orbits. Some astronomers thought Pluto was much bigger than the tiny dot implied and that most of the surface was hidden from view because it was dark. In fact, as it later turned out, the wobbles in the orbits of Uranus and Neptune were based on flawed observations, and there had never been any reason to search for planet X. Pluto's discovery was simply a coincidence.

It was not until 1950 that Pluto's size was measured, and the figure was confirmed 15 years later when Pluto was observed passing in front of a star. Around the same time, Pluto's day was noted as 6.39 Earth-days, which meant there were 14,100 Plutonian days in a Plutonian year

(248 Earth-years). Pale and dark patches seen on Pluto in the 1970s led to the theory that the poles might have more ice than the equator. But the real breakthrough in our understanding of Pluto came in 1976 with the discovery of methane in Pluto's spectrum. The ground had to be colder than –200 ºC (–328 ºF) for methane to exist as a solid. This also implied that Pluto was a small planet with a bright, frost-covered surface rather than a large planet with a dark surface.

Since the 1980s Pluto has passed in front of numerous stars, allowing astronomers to refine their measurements of its size and discover a very thin atmosphere, made mostly of nitrogen. Pluto is just 2324 km (1444 miles) in diameter – less wide than North America – yet it manages to hold on to a tenuous atmosphere. In fact, during Pluto's brief, 20-year summer, when nitrogen ice fields warmed by the sun turn to gas, Pluto's air might grow as dense as that of Mars, before it gets lost into space.

The Hubble space telescope was turned towards Pluto in 1996 and revealed a varied terrain, ranging from areas as dark as coal to paler zones as bright as snow. It is not known what the dark material is – some suggest primordial organic material or ancient ice that has darkened due to reactions with cosmic rays. Pluto's poles are brighter than the equator and probably have ice caps that expand and contract with the seasons, perhaps encroaching half way to the equator.

1 & 2. Pluto is little more than a tantalizingly blurred object even seen through the exquisite Hubble telescope.

Win Some, Lose Some

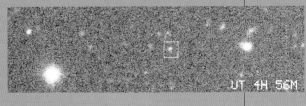

Almost half a century after Pluto's discovery, US astronomer James Christy noticed a lump on the planet. In Christy's images Pluto looked smudged – as though his telescope had been knocked while the photo was taken – but stars in the pictures remained sharp. Further investigation revealed that the lump moved around Pluto every 6.4 days. Pluto had a moon.

Charon, as the moon was christened, has a synchronous orbit around Pluto, which means that the pair present the same face to each other all the time. To an observer on the surface, Pluto and Charon would remain permanently fixed at the same position in each other's skies. Charon is very different from Pluto. It shows no evidence of an atmosphere or polar ice caps and its colour is more uniform, suggesting a surface predominantly covered by water ice rather than the exotic cocktail of nitrogen, methane and carbon monoxide that is thought to cover Pluto.

Charon is more than half as wide as Pluto, making it more of a second planet than a moon. But rather than considering Pluto and Charon to be a double planet system, many astronomers now question whether Pluto is a planet at all. As early as 1936, there were suggestions that Pluto was an escaped moon of Neptune. The discovery of Charon made that unlikely, but the debate over Pluto's planetary status has continued.

In the 1990s, following a technological leap in the sensitivity of telescopes, astronomers began to discover new objects moving through the outer solar system, even farther out than Pluto. The first ones were only a few hundred kilometres wide, but since 2001 astronomers have found several that are more than 1000 km (620 miles) wide. These were the first big members of the Kuiper belt – a long theorized region of icy objects thought to exist out beyond the planets. In 2004 the discovery of an object approaching the size of Pluto was announced. 'Sedna', as it was named, was dubbed the tenth planet, but many astronomers now suspect that Pluto, Charon and Sedna are all Kuiper Belt objects, rather than true planets.

1. Pluto and Charon as seen through ESA's Faint Object Camera on board Hubble.

2. The best view of Pluto and Charon from Earth.

3. Sedna – dubbed the tenth planet – so far away it takes 10,500 years to orbit the Sun.

4. Kuiper Belt Object 1995 QY9 caught moving against its star field.

5. Artist's impression of Sedna – the largest Kuiper Belt Object found so far is almost the size of Pluto, but it's three times further from the Sun.

Kirby and Grigorev construct the Lowell Telescope on the surface of Pluto. The eight separate mirrors will work as a single giant telescope capable of resolving Earth-sized planets orbiting other stars.

goes every 20 minutes or so. Nina and Zoë are busy mapping Pluto and Charon and are using the deep shadow on Pluto's far side to make observations of very faint objects far beyond our galaxy.

Earlier today, Yvan pointed out that despite our marathon voyage to the outer solar system at the very limits of human endurance, the constellations look exactly the same as from Earth. The stars feel no closer whatsoever. He said it was like we'd climbed Everest to get nearer to the Moon. Typical Yvan.

Nina Sulman, Mission Scientist
1388 days, 10 hours, 18 minutes

Pluto might mark the end of the road for us, but it also marks the beginning of a region of space that's populated by billions of fragments of rock and ice – leftovers from the formation of the solar system. These important relics from the birth of the planets, like Sedna, hold deep significance.

There might be nothing we'd class as life out here, but for me there is something as profound – the first organic molecules. The early solar system was probably awash with organic compounds – perhaps delivered on comets tumbling inwards from out here. They settled on the surfaces of the newly formed planets, and where conditions were right a few ended up forming structures that could copy each other. Occasional shoddy copying created ever more complex molecules, some of which began to flourish in groups. Groups became systems; systems became cells and cells became simple organisms. Eventually, complex multi-cellular organisms evolved; whose intricacy reflected the passing of unfathomable eons of time. They in turn designed and built their own complex machines, which carried them beyond the place they were born and out in search of their origins. It was a beautiful thought.

Nina Sulman, Mission Scientist
1394 days, 14 hours, 1 minute

The days we've spent in orbit around Pluto have been very special. I find this a peaceful and calming corner of the solar system, far from the deadly solar furnace and the lethal radiation belts of Jupiter. The frozen landscape below is beautiful to watch as it turns in the gentle light of the distant Sun.

Pluto's surface is ideal for the telescope. The planet is seismically stable, with no tidal gravity tugs or geological activity to shake the ground. In fact, Pluto is probably the most stable site in the solar system, and its low gravity makes building the telescope relatively easy. The low temperature also helps – it will chill our instruments to the point where they can detect the tiniest amount of heat emanating from distant worlds. And the trace atmosphere, just a few ten-thousandths of Earth's, will grow weaker and weaker as Pluto drifts further from the Sun. Over the coming years the telescope will move into Pluto's night, where the Sun's glare is hidden and the view is even clearer. On the far side of Pluto, powered by a small nuclear reactor, it will continue peering into the universe and bouncing data back to Earth off Charon's surface.

The telescope is not a single object but a circle of six mirrors, each about 2 m wide. The light from these mirrors is cleverly combined to make a single image, as though all six were parts of one gigantic mirror. The whole array was originally designed to probe the edges of the known universe, but we've reconfigured it to look for planets around other stars that might support life. Finding such planets is difficult, but this telescope will be a great help. It should have enough magnification to see planets many light years away, and careful analysis of the images will tell us whether those planets have the right kind of chemistry to support life. But before pointing the telescope into deep space, we decided to test it out on a closer target: Earth. We were eager to see what our home planet would look like from the farthest point in our mission, and Tom thought the telescope might give us an image of Earth as historic as those taken by the Apollo astronauts on the way to the moon in the 1960s.

So, after 12 days on Pluto, Yvan and Tom uncovered the mirrors and turned them towards the inner solar system. As we cleared the horizon on our 620th orbit and came into radio contact with Tom and Yvan, the picture started to appear on the monitors in *Pegasus*'s command centre. It wasn't the most beautiful or clearest picture of Earth from space, but it was somehow the most special. The tiny distorted image still reflected the uniqueness of our home world. I could see blue oceans gleaming in the sunlight, wisps of white cloud, and a single brown continent – probably Africa – a vision from the farthest point of human exploration of the continent from which the human race first emerged.

I was lost in thought when *Pegasus* dropped below the horizon again and we lost radio contact. The image broke up and vanished. I'd completely forgotten to test whether the spectrometer could detect oxygen in Earth's atmosphere.

✴ Tom Kirby, Commander
1395 days

The image of Earth came as a reminder. For three years we'd been travelling away from Earth, always looking outwards and always preoccupied with the next phase of the mission. Suddenly, here was an image of where we came from and where we truly belonged. Our work here at the edge of the planets was done, and it was time to head home. But before leaving Pluto, there was one last task to do.

'Earth is the cradle of humanity, but one cannot live in a cradle forever.' Not my words but those of the Russian visionary Konstantin Tsiolkovsky, written in 1911. Today, on our last day on Pluto, I laid a plaque inscribed with this quote next to the telescope. It was a fitting tribute to all the dreamers whose imaginations had propelled us here. Before placing the plaque on the ground, I read out Tsiolkovsky's quote and the inscription below it: 'We commend the bravery and commitment of all those brave souls who gave their lives to the exploration of space'. Above the text is our mission patch, and around it 32 stars; representing the astronauts and cosmonauts who had perished in our quest to reach for the stars.

Yvan had scratched an extra star into the surface for John. I laid it in the nitrogen snow.

The ceremony to mark the most distant outpost that humanity has reached.

Missions to the Edge

Pluto, Charon and the other Kuiper objects have probably changed little in billions of years. Barely touched by the Sun's heat, and too small to be geologically active, these bodies have probably been frozen solid since the days of the early solar system. This means they contain important clues to the formation of the inner planets, and frozen into them are the original chemical building blocks of the solar system and the life that has taken root within it.

Unfortunately, however, all are beyond reach of our best telescopes. Even the Hubble space telescope can see little more than a fuzzy, pixellated grey-and-white blob when it looks at Pluto. The only way to find out more about these icy worlds is to send a space probe.

NASA first planned their Pluto–Kuiper Express mission in the 1980s, but it was cancelled in the 1990s after rising costs. A call went out for more affordable designs, and the plan to explore Pluto resurfaced in the 'New Horizons Pluto–Kuiper Belt Mission'. Weighing just 445 kg (980 lbs), the *New Horizons* spacecraft should be launched in January 2006. It will fly first to Jupiter, where a gravity slingshot will throw it out of the plane of the solar system and accelerate it to 70,000 km/h (43,000 mph). Travelling in semi-hibernation to conserve power, it should arrive at Pluto in the summer of 2015.

If all goes to plan, the probe will 'wake up' when it is about 75 days and 100 million km (62 million miles) from Pluto. After flying for nearly a decade, it will pass within a few thousand kilometres of Pluto, map the surface of both Pluto and Charon at 1-kilometre resolution, and determine the chemical make-up of the surface ice. It will also try to find out what Pluto's air is made of and will search for any rings or smaller moons that might orbit the planet.

Unable to stop, *New Horizons* will have to accomplish all its work in a single flyby. With this phase of its mission complete, the spacecraft will reorientate itself and attempt one or more flybys of known Kuiper objects in the vicinity of Pluto over the following six years.

New Horizons cannot afford to slip behind schedule. If it fails to launch by 2007 it will miss the chance to slingshot around Jupiter, and the next launch window will not open until after 2020. By then, Pluto's southern pole will be in total darkness, hiding it from the cameras of a passing spacecraft. Pluto will also be much further from the Sun and in the grip of a 200-year chill that will have frozen its atmosphere. The planet's intriguing summer weather will not return until the twenty-third century.

1. The New Horizons Pluto-Kuiper Belt Mission flies on following its encounter with Pluto in 2015.
2. The Dutch born twentieth-century astronomer Gerard Kuiper. In 1951 he proposed a flattened ring of debris beyond Pluto as a source for short period comets.

Billions of Planets

Until just over 10 years ago, the nine planets of the solar system were the only ones we knew about. Of course, with all those stars out there, it was likely there were billions more, but how could we possibly find them so close to the glare of their stars? It turned out that large planets have just enough gravity to make their stars wobble a little,

and once astronomers figured out how to detect this tiny wobble, planets started popping up all over the place.

So far, around 120 'extrasolar' planets have been found. Most appear to be giant planets larger than Jupiter that hug their stars in very tight orbits. Whether this is typical of other planetary systems is hard to tell, but it could be a sampling error caused by the fact that only the biggest planets can be detected with present technology.

Today more than 1000 stars are currently being scrutinized for planet-induced wobbles, but if we are to find smaller, Earth-like planets we will need to improve our detection methods. And if we are going to find the chemical signs of life – water, oxygen, carbon dioxide and methane – we will need to catch a glimpse of the planets themselves.

Such an opportunity presented itself in 2001, when a planet slightly smaller than Jupiter passed across the disc of its star, HD209458, in the constellation Pegasus. This rare alignment allowed astronomers to get their first look at an 'exoplanetary' atmosphere. By analysing starlight passing through the edge of the planet, astronomers were able to record some basic chemistry. Subsequent observations made by the Hubble space telescope, announced in January 2004, have identified oxygen, hydrogen and carbon in the planet. It is the first world outside the solar system seen to contain some of the building blocks of life. Further observations will look for water vapour.

To probe the atmospheres of smaller, Earth-sized extrasolar planets, a new

breed of telescopes will have to be built. Such new telescopes would have to be interferometers, which work by combining light from several instruments to create a giant 'virtual telescope'. Astronomers already use ground-based interferometers to study objects in deep space. To study small extrasolar planets, we'd need to put interferometers into space.

There are already plans to fly such telescopes. One design, called the Space Interferometry Mission, will test the technology and pave the way for more ambitious missions, such as NASA's Terrestrial Planet Finder and ESA's Darwin Mission. Using a flotilla of space telescopes flying in formation, these giant virtual telescopes will be able to study the atmospheres of Earth-like planets up to 45 light years away.

Although the launch of the telescopes is more than a decade away, astronomers already have a shortlist of stars to look at. These first candidates, close to Earth and most like the Sun, should give us the best chance of finding a planet where life has evolved.

1. The star HD209458 in middle of this field is orbited by a planet with oxygen in its atmosphere.
2. The Earth photographed from 5.92 billion km away by *Voyager 1* in 1990.
3. The powerful Keck telescopes on Mauna Kea in Hawaii. Their twin 10-metre-wide mirrors work as if they were a single 85-m telescope.

Coming Home

EARTH

- ⊠ DIAMETER: **12,756 km (7972 MILES)** ⊠ MASS: **1 (5.9 X 10^{24} g)** ⊠ GRAVITY: 1 G
- ⊠ AVERAGE SURFACE TEMPERATURE: **14.4 °C (57.92 °F)** ⊠ DAYLENGTH (ROTATION PERIOD): **23 HOURS 56 MINUTES**
- ⊠ YEAR: **365.25 DAYS** ⊠ AVERAGE DISTANCE FROM SUN: **149.6 MILLION KM (93 MILLION MILES)** ⊠ RINGS: **0** ⊠ MOONS: **1** ⊠

After leaving Pluto, *Pegasus* set off on its return leg to Earth. The crew were still heroes in the eyes of the public, and some newspapers began a countdown to their arrival. And since *Pegasus* was carrying so much priceless planetary material, a few scientists at Control were probably ticking the days off too.

But there was still one mission goal left: a rendezvous with a comet. This had always been an optional extra because it relied on a comet appearing in the right place at the right time, just as the crew were heading home. As luck would have it, on day 1415 of the mission, a suitable comet appeared as a tiny speck on our telescopes. It was Comet Yano–Moore, a name that everyone on this planet now recognizes, though back then it was just another comet. I distinctly remember the anxiety I felt when I heard the news – after all we'd achieved, I felt we didn't need this. However, the scientist in me knew this was perhaps the biggest stroke of luck we had had all mission. The opportunity to land on a comet – a pristine, unspoilt fragment of the early solar system – was unprecedented.

Once we'd established Yano–Moore's trajectory, we held a big meeting at Control. The whole team was present. The atmosphere was tense, and it soon became apparent that the scientists weren't going to win the day – most of the team wanted the crew safely home. It was even agreed that we wouldn't mention the comet to the astronauts. This I did not like – we would have to lie, and that was something we hadn't done before. In the end it was all irrelevant. We hadn't counted on Zoë – a great astronomer with a telescope and time on her hands. Minutes after the meeting she popped up on screen, absolutely over the moon. She had spotted the comet, and the crew had decided that, if our calculations matched theirs, *Pegasus* should head for a rendezvous. Earlier in the trip we might have tried to overrule them, but things had changed and they were running the show now.

Zoë wanted to name the comet *Pegasus*, but she wasn't the first to see it. It had first been spotted by two amateur astronomers working independently in Japan and Europe, and so the name Yano–Moore stuck. The comet had originated a light year away, in the Oort cloud – a gigantic cloud of comet nuclei that orbit the Sun far beyond the planets. Comets from the Oort cloud are known as 'long-period comets' because their orbits last thousands of years and their visits to the inner solar system are few and far between. Yano–Moore had probably made only a couple of previous visits to the inner solar system, and that meant its surface should have been relatively stable.

The central portion of our own galaxy, the Milky Way.

Nevertheless we were unsure how safe it would be to land on. Comets become unstable and unpredictable as they tumble towards the Sun and warm up. The best time to visit, therefore, was while the comet was still in the outer solar system. At the same time, we wanted to minimize the communications delay with *Pegasus*. As a compromise, we opted for an encounter when *Pegasus* was slightly farther from the Sun than Saturn.

There was a backlash in the press, even though we made it absolutely clear that the crew themselves had made the decision. Naturally, it didn't take long for the media to find out how unpredictable comets can be and how little we know about their behaviour. But the die was cast and we couldn't order the crew to keep away from Yano–Moore – no matter how much I wished they would.

Alex Lloyd
Chief Scientist, *Pegasus* Mission

⊡ Tom Kirby, Commander
Mission elapsed time: 1402 days, 9 hours, 32 minutes

On the 1400th day of the mission, Yvan and I climbed on board *Clyde* for the final time and blasted off the surface of Pluto. The telescope was performing well and we had handed over control of it to Earth. Fittingly, the first planetary system they turned it on was in the constellation of *Pegasus*. I made a bet with Yvan that at least one Earth-like planet with an oxygen-rich atmosphere would be found before we made it back to Earth. He wasn't as optimistic.

We've both been resting since we returned to *Pegasus*. We worked extremely hard on Pluto, and I ended up feeling run-down. Now I've succumbed to some kind of cold and am feeling blocked up and groggy. I joked that I'd caught a cold on the coldest planet we've visited, but Nina pointed out that I could have contracted something from Pluto itself. I immediately became one of her experiments. For the first time in the mission, I found myself hoping that we hadn't discovered some form of extraterrestrial life. Nina almost looked disappointed when it turned out to be just a familiar virus.

While resting and getting over my cold, I've been sketching and painting Pluto's landscapes from memory. Throughout the mission I've been documenting our explorations in paintings and drawings. I find that waiting a few days before starting to draw helps me filter the images through my mind before I attempt to commit them to paper. As I finished the first painting of *Clyde* on Lowell plateau, I found myself wondering if people would ever return. Maybe I'll be the only artist ever to see that frozen world.

⊡ Tom Kirby, Commander
1415 days, 14 hours, 49 minutes

A long-period comet is usually discovered from Earth every month or so, and they seem to come from random directions. So we'd thought that the chance of finding one we could intercept was slim. But, just as we were running out of time to alter our return trajectory, Zoë spotted one. We wanted to call it Pegasus, but Control informed us that it had already been named Yano–Moore.

Nina is ecstatic. She's been hoping for a comet encounter every since we left Saturn, and now it looks like she's going to get one.

Comets come from the Oort Cloud. They are fossilized relics of the early solar system, preserving the chemicals that prevailed 4.6 billion years ago. Nina says the rendezvous will be 'the ultimate archaeological encounter with a mummified world'.

Control reckons we'll meet the comet in about 600 days, when we're about 1.6 billion km from the Sun. They think that at that distance the comet won't have started to shed material and should be safe to approach. We'll make a detailed visual inspection and, if it looks stable, Nina and Zoë will attempt to land on it and collect samples.

☒ Yvan Grigorev, Flight Engineer
1440 days, 8 hours, 35 minutes

We've now made the crucial engine burn to put us on course for the comet. It is a course that will also take us home, and with every day that passes we now get closer to Earth. We are now falling back into the Sun's gravity well rather than climbing out of it. The whole mood of the journey is different.

Doom and Gloom

EDMUNDUS HALLEIUS R.S.S.
Astronomus Regius et Geometriæ Professor Savilianus.

Chinese astronomers were the first to analyse the orbits and appearance of comets. They knew as early as 200 BC that a comet's tail always points away from the Sun, no matter which direction the comet is travelling in. They suggested that the tail's position was caused by solar 'energy', a remarkable insight that was proved correct more than 2000 years later when the solar wind was discovered.

Since ancient times, people have interpreted comets as bad omens, heralding epidemics, wars or other catastrophes. Pope Calixtus III reportedly considered the great comet of 1456 (later named Halley's comet) to be an agent of the devil. In Europe, such beliefs were holding back comet science and

they were thought to be atmospheric phenomena until an experiment conducted in 1577. The Danish astronomer Tycho Brahe observed a comet from various locations in Europe to see if it changed position against the background stars – but it didn't. So Brahe therefore concluded that it must be even farther away than the Moon.

It took some careful calculations to show just how far away comets were. In 1705 the English astronomer Edmund Halley studied past observations of 24 comets and concluded that they moved in elliptical (oval) orbits around the Sun. He also noticed that the comets of 1531, 1607 and 1682 had very similar orbits, and suggested they were in fact a single comet with a 76-year orbit. He

was proved right when it returned in 1758, 16 years after his death.

Despite these revelations, people continued to believe that comets were harbingers of doom. In 1773 the mistranslation of a mathematical paper sent Paris residents scurrying for safety in a church, thinking they were about to be struck by a comet. And as late as 1910, when Halley's Comet visited the inner solar system once more, anti-comet pills sold out in the USA.

We were a little wiser when Halley's Comet returned in 1986 to be met by an armada of spacecraft, including the European probe *Giotto*, which photographed the nucleus. Since its visit astronomers have continued to watch Halley through telescopes. In 1991, without warning, the famous comet suddenly flared up again. Quite what caused this unexpected outburst is a mystery.

In 2003, astronomers working at the Very Large Telescope in Paranal, Chile, managed to spot Halley's Comet again, now 4.2 billion km (2.6 billion miles) from the Sun – almost as far as Neptune – and totally inactive. The 10-kilometre (6-mile) wide cometary nucleus of ice and dust reflects just 4 per cent of the light falling on it and was nearly a billion times fainter than anything visible to the naked eye. It was the faintest comet ever seen. Telescope technology is now good enough to track Halley's Comet through its entire 76-year orbit – some compensation for the

fact that it won't be visible to the naked eye again until 2061.

Halley's Comet is a predictable visitor to the inner solar system, but many comets appear out of the blue and are never seen again. One of the most impressive of these unexpected visitors was Comet Hale-Bopp, discovered simultaneously in 1997 by two amateur US astronomers, Alan Hale and Thomas Bopp. Dubbed the comet of the century, it lit up night skies across the world with a pair of dazzling tails that grew to 100 million km (62 million miles) in length.

However, I'm not entirely thrilled at the prospect of exploring Comet Yano–Moore. Maybe I should be more excited, but I'm tired. I feel our tour of duty has done enough. *Pegasus* has been in space almost four years, and we still have a couple of years left to fly. With little to do but watch the Sun get brighter, I'd hoped this last leg of the voyage would be quieter. But now there's the comet lander, *Messier*, to prepare and new simulations to programme. And the prospect of approaching a potentially unstable comet does not fill me with joy.

🎞 Nina Sulman, Mission Scientist
1480 days, 19 hours, 45 minutes

Suddenly we've something new to train for – a comet encounter. I feel re-energized. I've studied comets for as long as I can remember, but I never dared dream I'd get to visit one. I'm hoping it will be stable enough to attempt a series of landings in *Messier*. I've already started training in the simulator for my role as copilot. Comet Yano–Moore has travelled 6 trillion km to meet us, and I'm determined not to screw up.

The descent to the surface will be more like a docking manoeuvre than a landing. The gravity of comets is so weak that we're more likely to bounce off the surface than fall towards it. *Messier* has specially chilled harpoon bolts that will anchor us to the ice. Even the footpads are specially chilled so they won't melt into the ground. Comets are pretty cold places – just ten degrees above absolute zero, the lowest temperature possible. A hot spacecraft, engine exhaust, harpoon bolt or even a sun-warmed ladder leg might slice though the ground like a knife through butter – or, as a colleague once said, like a marble through a milkshake.

With *Pegasus* in a safe parking orbit, we'll make the first of many excursions in *Messier*. With so little gravity to contend with, the lander will be able to fly back and forth to the comet without using much fuel. Down on the surface, our EVAs will probably feel more like scuba dives than spacewalks, as we drift over the terrain with our jet-propelled backpacks to guide us.

Time will be tight. The closer a comet gets to the Sun, the more unstable it becomes. Eventually the solid part of the comet – the nucleus – will start to disintegrate, spewing out dust and gas as carbon monoxide and carbon dioxide boil off the surface. These emissions form the fuzzy white cloud, or coma, and the long bright tails that give comets their unmistakeable appearance. Quite when this disintegration begins depends on an individual comet's chemistry. And until we get a better idea of our comet's chemistry we won't know how unstable it is likely to become.

It's risky. Comets have been seen to break up without warning. In 1991, while heading away from the Sun, Halley's Comet suddenly just threw out a huge dust cloud. We still don't know why it did this. It could have been a collision with something. Or maybe a pocket of carbon dioxide suddenly burst. If Yano–Moore has any surprises in store for us, at least *Pegasus* won't be far away. We've pulled off some very challenging missions, and I'm confident we can do this one too.

After all, we've got more than a year to practise.

👁 Zoë Lessard, Mission Scientist
1556 days, 15 hours, 13 minutes

There's another long slog ahead of us now, with very little to do but wait. There's the comet encounter to prepare for, but it still feels like a hell of a long way away. I was excited at first, but now I realize what a mountain it will be to climb. I know Nina is really up for it, but the others seem to be less motivated than at any time in the mission. Food has taken on more significance for Tom and Yvan. Tom is always in the mess area when I get there, and he lingers for ages after Nina and I have finished, tucking into extra helpings with Yvan and appearing reluctant to return to chores.

Flicking through my digital image collection the other day, it occurred to me that I haven't used the words 'car' or 'train' for probably three and a half years. Strange thought, I know, but when I began thinking about it there was a whole bunch of words we don't need up here. 'Wind', 'rain', 'cat', 'dog', etc. There's just no reason to talk about such things. It all adds to my feelings of detachment from Earth.

For relaxation I sometimes watch footage of natural environments on Earth – forests, lakes, mountains, and so on. For extra realism I wear a virtual-reality headset and listen to stereo sound effects at the same time. But these scenes seem increasingly archaic and unfamiliar, perhaps even alien. The smells, tastes and feelings associated with them are such distant memories. We have a set of 'aroma pouches' to go with the views,

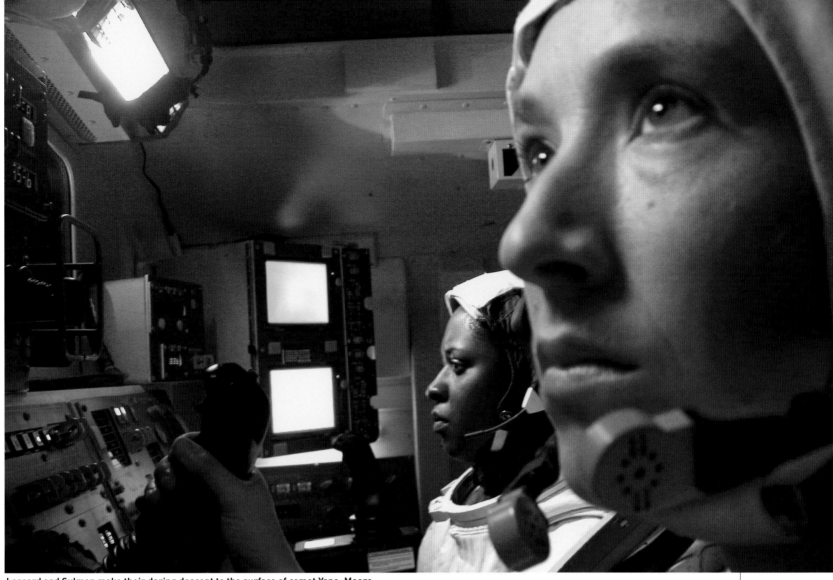

Lessard and Sulman make their daring descent to the surface of comet Yano-Moore.

and they do help to transport us away from the metallic smell of the spacecraft. But I long for the real thing.

 Zoë Lessard, Mission Scientist
1643 days, 14 hours, 22 minutes

Yvan has taken on the ship computer at chess again. He used to play in the early months of the mission, but he gave up after never winning a game. Now he's taken it up again with renewed ferocity and spends hours in his sleep module trying to beat it. He also seems to be working his way through the entertainment computer's entire MP3 collection. I'm glad the sleep compartments are well soundproofed!

I've found that I prefer to float in the storeroom reading novels. It's the biggest space in the ship and I can stay there undisturbed, without feeling trapped as I often do in the half gravity of my claustrophobic sleep area. I've also neglected my daily exercise

programme, lacking the motivation I had earlier in the mission. I know it's detrimental to my health, but I guess I've given up caring. And without John to keep a watchful eye on our muscle and bone strength, it's easy to shirk those responsibilities. And Nina is too busy fantasizing about the comet to even notice.

⊠ **Yvan Grigorev, Flight Engineer**
1921 days, 12 hours, 59 minutes

Since Zoë discovered it, Yano-Moore has been rather unspectacular – just a clump of dark pixels on a slightly darker background. But now it's getting big enough for us to make out a few features. The surface looks pockmarked, and we can see it slowly change as the comet tumbles around. It all looks slightly blurred, even through the best telescope. Nina tells me this is because carbon dioxide is already leaking out and throwing bits of dust off with it. It has probably been doing that for years, she says. But it isn't a reason to be alarmed.

That's all right for her to say. She hasn't spent six months on spacewalks repairing an aeroshield cratered with impacts from 'bits of dust'. Sure, the comet still looks small and harmless through the telescope (especially from this distance), but appearances can be deceptive. I know it has the potential to suddenly explode without warning. It could prove to be more than *Pegasus* can cope with – and we'll still be a billion miles from Earth, with no hope of a rescue. If things go badly wrong, we're dead.

◉ Zoë Lessard, Mission Scientist
2006 days, 10 hours, 20 minutes

Here we are in orbit around a city-sized hunk of black ice. After so long with nothing but stars to see, it feels strange to be flying so close to this pocket-sized world. It spins quite fast about its long axis – completing a day in just 12 hours – and turning slightly slower about its short axis. The two axes of rotation made getting *Pegasus* into orbit tricky, but we're relieved the comet didn't have a more complicated motion.

Barely caught in the low gravity field, *Pegasus* orbits Yano–Moore at a sedate speed – not much faster than a cyclist. This hunk of rock looks like it should weigh more pulling us around a bit faster, but it has a loose, rubbly interior that is actually less dense than polystyrene. And it's getting lighter by the day as it sheds matter.

Since reaching orbit we've kept a round-the-clock vigil. Yvan pushed for it. As he points out, there's every possibility the comet could break up suddenly. There's also the possibility that newly emerged jets of gas could push it off course and towards *Pegasus*. This close, our collision avoidance system might not provide enough warning to keep us safe. There's no better warning system than eyes and brains, so we take it in turns to keep watch. It isn't wasted time – there's an extensive mapping programme to complete before Nina and I can even think of venturing down there.

Our main task at the moment is to pick a series of landing sites that look both scientifically interesting and safe. But that's easier said than done. There don't seem to be any flat areas, and a fine fog clouds our view of the dayside. We're employing a battery of

instruments to examine every chemical and physical property the science team can think of. We're using radar to probe the comet's interior and locate water ice; irregularities in our orbital speed are translated into gravity maps; and spectral analysis of the surface will reveal the distribution of organic molecules.

There are two schools of thought on the landing sites. Half the scientists at Control think we should land on the dayside, which is well lit but misty. The other half think we should make an instrument-based landing on the nightside, away from the glare of the Sun and the mist. Either way, we'll be partially blinded.

In the end, they left it to us to decide – and I've never been one for night flying.

◉ Zoë Lessard, Mission Scientist
2016 days, 12 hours, 30 minutes

Ten days of mapping turned up six possible landing sites on this tumbling mountain, and today we headed out to the first one. Nina and I powered up *Messier* and completed our pre-flight checks while *Pegasus* was on the nightside of the comet. As soon as the Sun rose, Tom gave us the all clear and I flipped the release switch. *Messier* slowly detached from the docking port and moved into the sunshine. It's been a long time since I left *Pegasus* and I felt uneasy at first, conditioned almost – I suppose I've become a bit agoraphobic. I looked at Nina for reassurance but she had one of her determined looks on her face. I pulled myself together and focused. I couldn't afford to be distracted on a descent like this.

Unlike the other landers, *Messier* can't rely on gravity to pull it to the surface, so it has an extra set of rockets on the roof to propel itself down. I fired them to begin our descent. It felt more like controlling an MMU than a spacecraft. The hardest part of the descent was matching our flight to the two-axis tumble of the comet, while simultaneously trying to bring *Messier* steadily towards the landing site. Computers handled the former; I handled the latter. Constant puffs and bursts of our positioning engines, fired by the computer, countered my own actions. Responding to two pilots in this way, *Messier* pitched and yawed as we powered down to the surface.

It was extremely rugged down there, with undulating plains rising into steep, almost vertical slopes. Everywhere the ground was peppered with overlapping craters and streaks of debris thrown out by impacts.

Messier **drifts just above the comet's surface –**
tethered in place as Lessard and Sulman start their EVA.

Nina counted down our height and read out our horizontal position. I kept my eyes on the surface and searched for a spot flat enough to place the footpads. Because of the comet's small size, the horizon was never far away, and each time *Messier* cleared a hill, a brand new landscape opened up beneath us.

As we neared the surface, a little too fast, I ignited the main engine beneath the legs in an attempt to slow our descent. Billowing clouds of dark dust surged out around us, splattering the windows with sooty flecks. I'd slowed the descent to less than half a metre per second as Nina's altitude count reached single figures. But we still bounced as we made first contact with the surface. I shut down the engines immediately and fired the harpoons. The thumps broke the silence as the three rocket-propelled bolts punched deep into the ice, attaching firmly as chilling mechanisms froze them into the ground. 'Capture,' shouted Nina. It felt like we were clinging onto the surface by our fingernails.

🔲 Nina Sulman, Mission Scientist
2016 days, 21 hours, 50 minutes

We're both filthy, covered in precious, sticky hydrocarbon comet chemicals. What an extraordinary place this is. It's exhilarating being here, I feel invigorated by the energy this place seems to have. The ground is constantly fizzing in the sunlight as it exhales its murky breaths. It is almost as if it's alive.

We drifted out of *Messier*'s hatch about an hour after landing, and there we hung – floating above the surface, hardly touched by the low gravity. Even fully suited and strapped into a bulky MMU, I barely weighed more than a pencil on Earth. It took a full two minutes to drop the 3 m to the surface, and my feet touched down almost without me noticing. I'd been distracted while I was guiding Zoë out, and suddenly I glanced down to find I was 'standing' on the surface. And what a surface it is. Sinisterly black – darkened by billions of years of cosmic ray bombardment, and so rough and fragile. It reminds me of the blocky, jumbled lava piles you find in Hawaii.

It's impossible to walk – the slightest step sends you flying upwards. Instead, we use the MMUs to manoeuvre like scuba divers swimming across the comet's floor. Irregularities in the weak gravity field can still tug you unpredictably though, and it's sometimes necessary to tether yourself to the surface when you

want to remain still. I lost a couple of samples and a coring tool when I knocked them with my suit – they went spinning off with enough speed to carry them into orbit.

Much of the surface is badly eroded and pitted. Where the ice has boiled off, it has left intricate structures in the surface – tiny, twisted strands of black material. In the sunlight it's sticky, a bit like tar. Once you touch it you can't get it off. We bagged up some samples near *Messier* and then ventured farther, working together to find a fresher, less corroded surface to study.

We figured that our best bet of finding fresh material was to look in a crevasse, where the surface is still hidden from the Sun. Radar mapping from orbit had detected a promising area a few hundred metres 'south' of *Messier*, so we set off there with a burst of acceleration from the MMUs. I had to make repeated tiny adjustments to keep from bumping into the surface. It was like beachcombing – constantly inspecting the ground for anything interesting.

After about 15 minutes we reached a region that seemed more fractured. Narrow troughs began to appear, their depths hidden in shadow. They offered just the right sort of protection for potentially fresh material, but the first ones were too narrow to get even a sampling tool down. We flew on and found wider fractures, including one that was big enough to enter. Zoë went in feet first, examining the ice in front of her as I guided her down with my headlamp. It was dangerous and she risked getting stuck, so she moved slowly and stayed well within the beam of my headlamp.

As she slipped out of the sunlight, her suit sensors detected a massive drop in temperature, from 50 °C at the surface to −180 °C only a few metres deeper. She kept up a running commentary: 'It's so honeycombed … more cavities than rock and ice … it's less sticky in here'. Every small observation was a priceless piece of information that we only gathered because of the extraordinary effort and luck that had brought us this far. I was absolutely living every moment.

Back in *Messier*, we repressurized the cabin and got out of our suits. As soon as I removed my helmet, the smell hit me. The comet material on our suits had warmed up in the cabin and was vaporizing. It reminded me of coal tar soap.

We're now taking our scheduled rest period and hope to get a few hours' sleep before the next EVA. Looking out of the windows, I've noticed for the first time how regularly gas vents are breaking the

Lessard caught off guard by the comet's irregular gravity field as she manoeuvres her MMU jet pack towards a location to sample.

Lessard surveys the damage to *Pegasus* out of *Messier's* window.

surface and spraying out the dust that is starting to form a coma. The dust mostly looks as fine as smoke, but here and there are occasional fluffy 'snowflakes' a few millimetres in size. There's none of the geysers or jets you see in movies – just a slow drift of vapour, like mist rising from a morning meadow.

Nina Sulman, Mission Scientist
2017 days, 6 hours, 5 minutes

We'd only been asleep a few hours when a loud rattling noise woke us up. It was a tremor and *Messier* was shaking. It didn't last very long. I thought it was probably just a pocket of gas vaporizing on the dayside, so I tried to go back to sleep. But then there were more tremors. Now Zoë's upset – she says it reminds her of Io. She has that shell-shocked look about her again, something I haven't seen since Jupiter.

I've also noticed that the coma looks thicker, and the view from the window is not as clear as it was. A drop in visibility would make our EVAs much harder and might even mean a mission abort. Zoë's talking to Tom. He also detected the tremors, and he says they had to correct *Pegasus*'s orbit while we were sleeping because the comet's trajectory changed slightly, probably because of a gas eruption on the dayside. There's definitely something strange happening to Yano–Moore. Zoë wants to

abort. Tom is hesitant. He sees no reason to quit just yet – but advises we keep a careful eye on the tremors. We're going to suit up in preparation for the next EVA.

▨ Alex Lloyd, Chief Scientist
2019 days, 9 hours, 8 minutes (personal diary entry)

God, oh God, why did we let them get so close to that comet, let alone land on it? I've been standing outside, staring up at the night sky and willing them to respond to our continued requests for contact. We haven't received a signal for two days now. Everything had been going so well. Zoë and Nina's first EVA was a complete success, and we just expected the successes to carry on. I even checked the quake data and agreed with Tom that it was safe to continue. Then, in a split second, everything changed. Something has clearly gone badly wrong. I've lost count of the number of crisis meetings I've been in. No-one is giving up hope, but the telescope views of Yano–Moore are shocking. The comet brightened markedly at the same time we last heard from the crew. It must have broken up, perhaps taking our mission with it.

☒ Yvan Grigorev, Flight Engineer
2019 days, 18 hours, 30 minutes

Well I have nothing better to do now than write this report. I complain loudly about wanting to help, but I know the pain would make me worse than useless.

Looking back, we were given plenty of warning signs. Even when we first arrived, I suppose the comet was telling us to go away. By the time we had to move *Pegasus* when Nina and Zoë were asleep, I was convinced we should abort, but Tom overruled me. So when the nucleus did break up, *Messier* was still down on the surface. I was in the command centre and Tom was watching in horror from the cupola. I thought *Pegasus* was still safe. But I was wrong.

It was too quick to know what was going on at first. There was a blinding flash that lit up the whole cabin and a deafening bang. I felt as if someone had shoved me in the back. As I tried to collect my thoughts I became aware of a searing pain in my left shoulder

Messier **attempts an emergency launch from the surface of comet Yano–Moore as it breaks up catastrophically.**

A Comet's Tail

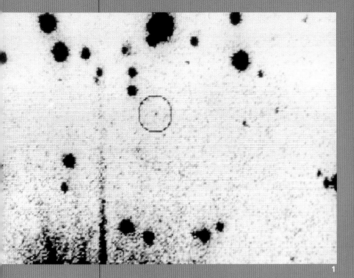

Until it gets close to the Sun and starts to spew out gas and dust, a comet looks much like an asteroid – an irregular chunk of ice, dust and rock that can range in size from a few kilometres wide to 40 km (25 miles) wide (the size of record-breaking Comet Hale–Bopp). This solid part of the comet is called the nucleus. The chemical composition of comet nuclei has been established by observing the gases they release. They seem to be at least 50 per cent ice – mostly water ice, frozen carbon monoxide and frozen carbon dioxide. The proportions of these vary widely between comets, and there are also small amounts of carbon-compounds and silicate minerals.

As a comet nears the Sun, the ice vaporizes and forms an atmosphere, or coma, that grows in size and brightness. The coma and nucleus together form the head of the comet. As they near the Sun the solar wind blows the gas and dust away from the comet, forming a tail millions of kilometres long that reflects sunlight and fluoresces in ultraviolet light. Dust particles tend to form a curved tail, while lighter material, such as charged atoms, forms a thick straight blue tail directly behind the comet. As a result, bright comets have two tails.

Not all the debris that is thrown off a comet leaves it for good. Large fragments sometimes tumble back onto the surface, covering the nucleus with rubble and stifling the development of a coma. In this way a comet might end up looking just like an asteroid for a period, until the ice layers are exposed once more.

Knowledge of the interior structure of comets is sketchy. What little we know is based on comets that have broken apart while being observed. These break-ups were apparently triggered by relatively small forces, which suggests the nucleus is loose and unstable, more like a pile of rubble than a solid block of ice.

1. The first sighting of comet Halley on its 1980s return to the inner solar system.
2. Comet Hale–Bopp in 1997, with its bright dust tale and a fainter blue plasma tail.

Missions to Comets

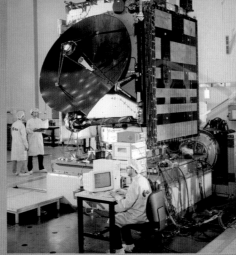

The first space probe to have a close encounter with a comet was the *International Sun-Earth Explorer*, which flew past Comet Giacobini–Zinner in 1985. The probe was originally built to study the solar wind, but it was diverted to intercept Giacobini–Zinner and renamed *International Cometary Explorer (I.C.E.)*.

The following year was a big one for comet missions – it was the year Halley's Comet returned. Russia sent its *Vega 1* and *Vega 2* probes; Europe sent *Giotto*; and Japan sent *Suisei* and *Sakigate*. It was a once-in-a-lifetime opportunity, and the probes bristled with a wealth of sophisticated instruments, from cameras and spectrometers to particle detectors and magnetometers. *Giotto* was sent up behind the comet and returned spectacular images of the nucleus before a piece of debris knocked it off course just 600 km (370 miles) away. The peanut-shaped nucleus was 16 km (10 miles) long and surprisingly dark, but just as predicted it was mostly ice. Geological features included a massive crater, some

180 m (600 ft) deep, and a hill of even greater proportions. *Giotto*'s cameras were damaged, but enough instruments survived for it to make a second encounter with a comet, called Grigg–Skjellerup, in 1992.

Another rash of cometary missions took place at the turn of the century. In 2001 NASA sent *Deep Space 1* on a successful flight past comet Borelly. And a probe called *Stardust*, launched in February 1999, flew through the tail of comet Wild 2 in January 2004, snatching samples of comet dust in a special catching device. These samples will parachute back into the Arizona desert in 2006. Another ambitious comet probe, called *CONTOUR*, launched in summer 2002, but it broke apart in Earth orbit when its rocket malfunctioned. NASA's *Deep Impact* is scheduled for launch in December 2004 and will visit comet Tempel-1 in 2005. Its mission is straight out of a Hollywood blockbuster: it will attempt to shoot a massive block of copper into the nucleus and blow the comet apart.

Also launched in 2004, Europe's *Rosetta*

probe will attempt the first landing on a comet. *Rosetta* was initially designed to land on comet 46P/Wirtanen, but delays forced a change of target to a much larger comet called 67P/Churyumov–Gerasimenko. The historic touchdown, by a lander called *Philae*, is planned for late 2014 on the summer side of the nucleus, when it is brightest. The lander will self-eject from the *Rosetta* orbiter at just 1 km (0.6 miles) from the nucleus. Touchdown will take place at walking speed, and harpoons will anchor the lander to the ground to prevent it bouncing off in the low gravity. *Philae*'s instruments will study the composition and structure of the surface and watch how conditions change during the short day–night cycle. A drill will extract samples from up to 30 cm (12 in) deep. The lander is designed to operate for at least 65 hours, but it might continue working for months.

The *Rosetta* orbiter is equipped with cameras, spectrometers and other instruments that will provide, among other things, very high-resolution images and information about the shape, density, temperature, and chemical composition of the comet. Hopefully, the mission will help scientists determine whether comets could have seeded the primordial Earth with the organic chemicals and water that made life possible.

1. Comet Wild 2 photographed by NASA's *Stardust* spacecraft in 2004.
2. The *Rosetta* spacecraft and its lander *Philae* undergoing testing before launch.
3. Halley's Comet photographed at less than 600 km away by ESA's *Giotto* spacecraft in 1986.

Comet Jano–Moore starts to develop a spectacular pair of tails as it journeys in towards the Sun.

and heard a loud gushing noise. A side window was punctured. The glass was holding, but a pinprick hole was venting air into space. The pressure alarm went off, adding to the confusion. I rushed instinctively for the portable oxygen reserve, but when I tried to move I realized I was badly injured. To compound my problems, I noticed that smoke was pouring out from behind the consoles. As we now know, a small piece of high-velocity comet debris had smashed through the cabin window, travelled though my shoulder and buried itself in one of the consoles, triggering a fire.

I managed to reach the oxygen supply, but my attempts to control the fire were pathetic. For one moment I thought I'd had it, but then Tom was next to me, wearing his gas mask. He activated the fire extinguishers, which quickly pumped the panels full of suffocating foam.

With the command centre fire under control, we floated to the data screens to look at a summary of damage to the ship. We'd been caught in a cloud of high-speed debris that had peppered the habitation module with impacts. The skin of the spacecraft is usually quite good at protecting against punctures, using a special layer of tiny, closely packed balls that automatically plug

holes as they are sucked out into space. But the comet debris had ruptured this layer and air was gushing out. The damage had also shut down a number of critical systems, including our main communications hub.

Not only did we fear losing *Pegasus*, but we were sure by now that *Messier* was already gone. Even if Zoë and Nina could successfully navigate between the biggest fragments, the cloud of smaller debris seemed certain to destroy them.

◉ Zoë Lessard, Mission Scientist
2019 days, 19 hours, 25 minutes

Debris pinged off the shield above us like hail on a tin roof. I couldn't see anything out of the window – visibility was down to metres. It was terrifying. Before the dusty coma had blotted out our view, I saw the horizon break apart. We both knew the nucleus had disintegrated. We just didn't know how many fragments there were, or whether we'd crash into something as we lifted off. But we sure as hell couldn't stay there, so we just throttled up the main engine and flew blind. Nina diligently counted up our progress as radar altimeters intermittently reported our height above what was left of the landing site. At any moment I expected *Messier* to crumple like a drinks can as we slammed into an unseen fragment of nucleus. It all seemed to go on for an eternity.

The pinging sound rang in our ears as we climbed, and then it suddenly stopped. We'd cleared the coma. At the sight of a bit of black sky and some stars, we both yelled with relief. But then we quickly realized that we had no idea where *Pegasus* was. I tried to reach Tom on the radio but the airways were dead. We assumed there was some serious damage to our antenna.

We tried not to panic. Nina started to work out where *Pegasus* should be, based on the most recent data we had. It was really fortunate that *Messier* had spare fuel and was so manoeuvrable. After two of the longest hours of my life, we got a visual on *Pegasus*, and after another 80 minutes we'd caught up with her. That's when we got the next big shock. As we drew close it became obvious that something was seriously wrong. The command centre was splattered with black material, and gas was venting into space from tiny holes in the hull. Still unable to contact Tom or Yvan, we decided to fly around *Pegasus* and alert them with our presence. That didn't work either, so Nina suggested we EVA across to the airlock and let ourselves in. We still had the MMUs and were close enough for this to be quite a routine manoeuvre. There was no point staying in *Messier* – if *Pegasus* was beyond hope, then so were we. The sooner we got on board and assessed the damage, the better. At that stage I didn't want to think what had happened to Tom and Yvan.

Within an hour we were back on board and confronted by the full horror of the damage that Yano–Moore had inflicted on our home.

⭐ **Tom Kirby, Commander**
2020 days, 11 hours, 40 minutes

It has been an emotional few days. The sight of Nina and Zoë emerging from the airlock was a rare moment of joy for me. We'd given them up for dead and were struggling to keep ourselves alive. Their return from beyond the grave was a huge boost, and

they made a big difference in our fight to save *Pegasus*. Yvan is in a bad way – one of his lungs collapsed – and with him out of action, the rest of us have had to work around the clock to patch the holes in the hull and make the most urgent repairs.

It was also great hearing back from Control. The return transmission we picked up three hours after our initial communication was jubilant. They offered us all the support they could muster, and they talked Nina through a chest drain procedure to stabilize Yvan's condition.

We turned John's sleep module into an operating room, and I'm pleased to report that the procedure went well. Much to his irritation, I've ordered Yvan to stay in bed and rest. He's finding it hard to lie there watching movies and listening to music with so much to repair on *Pegasus*, but he doesn't have much choice. He's useless to the rest of us at the moment.

We moved *Pegasus* a safe distance from the comet, upstream of the growing tail of dust and gas that now stretches behind the bright nucleus. There are few places in the solar system that I hope humans will never visit again, but this is one of them. Even if we wanted to return, there is little of Comet Yano–Moore's nucleus left to land on. And in any case, Control tells us that its trajectory has changed – a year or so from now it will slam into the Sun and cease to exist.

Now that *Pegasus* is habitable, our attention has shifted to the last planet we'll visit. Godspeed to us all.

Flight Surgeon Claire Granier monitors the operation on Yvan Grigorev.

A Comet Nursery

Astronomers divide comets into three groups, based on how long they take to orbit the Sun. The 'Jupiter family' of comets come round in less than 20 years and always appear within 40 degrees of the plane of the solar system, which suggests they originate in the Kuiper belt. 'Intermediate-period' comets have orbits lasting 20–200 years; and 'long-period' comets, which come from random directions in the sky, have orbits that can last millions of years.

Long-period comets seemed to come from outside the solar system. But in 1950, a Dutch astronomer called Jan Oort suggested they might come from a vast, spherical cloud of icy bodies that surrounds the solar system. This giant reservoir of comets, called the Oort cloud, has never been seen, and there's no direct evidence that it exists. However, the idea neatly explains the behaviour of the many comets that have orbits lasting a million years or more.

Oort reasoned that the icy bodies in the cloud could be disturbed by the gravity of passing stars, which would send them either hurtling into interstellar space or diving into the solar system. Such a stellar disturbance seems to happen every million years or so. It is thought the next star to disturb the Oort cloud will be a small red dwarf called Gliese 710. It could trigger an estimated 25 per cent rise in the frequency of comets about 1.4 million years in the future.

Every 30 million years, when the solar system crosses the plane of the Milky Way galaxy, the gravity of passing stars might trigger a dramatic shower of comets from the Oort cloud, increasing the frequency of comets by up to 300 times. This could account for the periodic mass extinctions that pepper Earth's fossil record and often coincide with raised levels of a rare, extraterrestrial material in the rock strata.

1. Professor Jan Oort at home.
2. A computer-generated image of the Oort cloud.

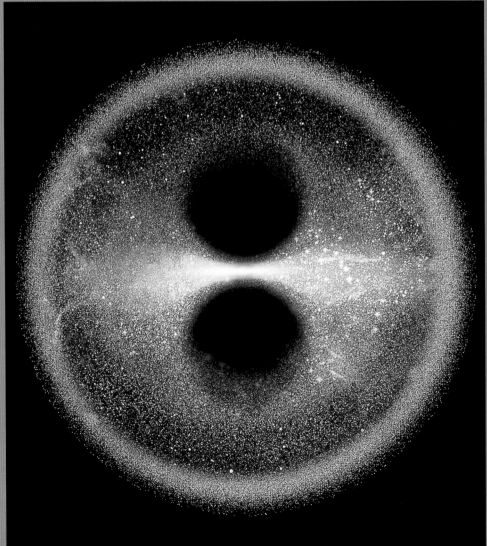

Uranus and Neptune

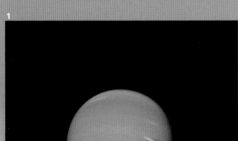

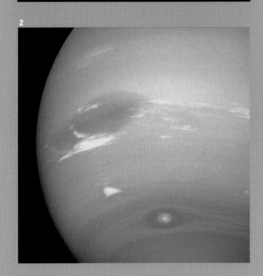

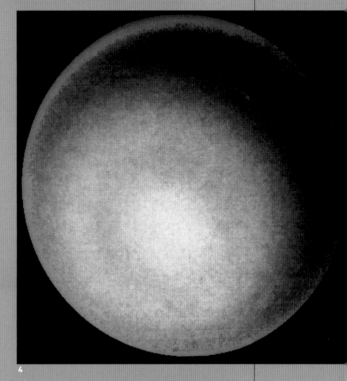

Beyond the reach of the *Pegasus* mission in our story, the outer ice-giant planets Uranus and Neptune take 84 and 165 years to orbit the Sun respectively.

The German-born British astronomer William Herschel discovered Uranus in 1781. It is four times wider than Earth and, despite its serene appearance, has a violent history. A massive collision tipped this giant world over onto its side, so it rolls around the solar system rather than spinning more upright as most of the other planets do. Uranus turns on its axis in 17 hours and 14 minutes, but the extreme tilt gives it peculiar seasons and days, with each pole experiencing 42 years of daylight followed by 42 years of darkness. The planet's dark moons also bear witness to a turbulent history – the moon Miranda looks like it was smashed apart and then reformed from fragments. Another moon seems to have been destroyed completely, leaving a series of thin debris rings around the planet. Uranus gets its bluish-green colour from methane clouds in the upper atmosphere, which absorb red light but reflect blue and green. Below the visible clouds is an atmosphere of hydrogen, helium and methane, and below that is thought to be an ocean of water ice, methane and ammonia. A rocky core is believed to lie in the centre.

Neptune is the outermost of the giant planets and was discovered in 1846, in a similar way to Pluto – after its existence was inferred by wobbles in the motion of Uranus. Slightly smaller than Uranus, Neptune also has a series of faint rings, discovered by *Voyager 2* when it flew past in 1989; the outermost ring is incomplete and consists of a series of arcs. *Voyager* found that Neptune and Uranus have stormy skies – a surprise in planets that are so far from the Sun. Neptune's winds are the fastest in the solar system. Giant, dark cyclonic storms seem to form in the clouds, but are shorter lived than those on Jupiter. Neptune's bluish colour is also caused by methane in the upper atmosphere, and the planet's interior is probably similar to that of Uranus, with a mantle of icy water, methane and ammonia overlying a rocky core. Of Neptune's eight known moons, Triton is the most intriguing. It is larger than Pluto and probably chemically similar, with a surface of nitrogen and methane frost. *Voyager 2* photographed its southern hemisphere and, to everyone's astonishment, revealed dark streaks of material sprayed across the surface by geysers. No-one had expected to see geological activity on this tiny, frozen moon. The solar system is full of surprises.

1. An enhanced picture of Neptune shows a wispy 'cirrus type' cloud overlying the Great Dark Spot.
2. Neptune's Great Dark Spot. Below is a bright feature that was nicknamed 'Scooter.'
3. A Hubble telescope picture of Uranus revealing the rings and some of the moons.
4. This computer-enhanced *Voyager 2* image of Uranus shows haze in the upper atmosphere.

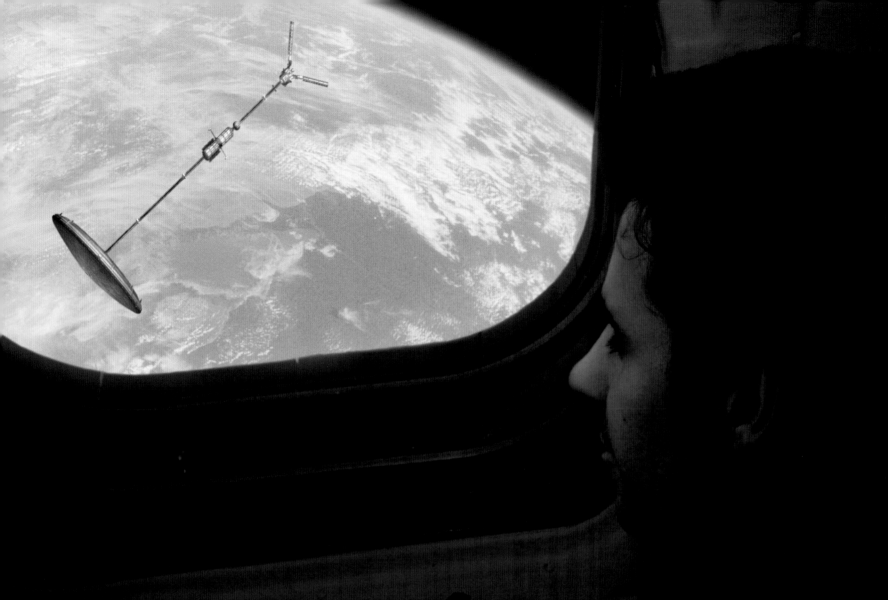

Pegasus returns to Earth orbit after over six years away from home, to be met by the support team.

🔖 Nina Sulman, Mission Scientist
2220 days, 13 hours, 22 minutes

I've always wondered how I'd feel returning to Earth after such a long time away. That glimpse of it through the powerful Pluto telescope more than two years ago reminded me what a beautiful, blue, marbled world it is. We've visited some very bleak and inhospitable places in this grand tour, hunting for a sign that life might exist beyond the Earth. But we found next to nothing. And now, as I look back at Earth, only 20 days' flight away, it feels all the more special because of that. We are coming home, and it is a precious home to come back to.

☒ Yvan Grigorev, Flight Engineer
2235 days, 11 hours, 12 minutes

Six years in space, and I'm determined to walk out of the capsule when we get back to Earth. Even if I never walk unaided again, I'm determined to walk away from the crew-return vehicle and into the quarantine centre. It's an important statement to all those watching that humans can leave Earth and visit other worlds. For the first time in history, we've shown that we can live for years beyond the Earth. We've carried our own portable biosphere around the solar system and back home, keeping healthy enough to walk on Earth once more.

Chapter 7

◎ Zoë Lessard, Mission Scientist
2241 days, 12 hours, 5 minutes

We reached Earth orbit just a day ago and, as if visiting another new world, we've been crowding into the cupola to soak up the view. Anyone would think we'd never seen it before. And I guess we never have with these eyes, which have seen so much of our planetary system and which now look afresh on our home – the Earth.

I worry a little about getting back into life on Earth after so long on *Pegasus*. Control have told us we can stay in orbit a few days or even a week before going back down. They'll send up a team to welcome us in another day or so, but even that's going to be pretty weird. I've not met anyone apart from Tom, Nina, John and Yvan in six years, and I don't know how I'll react to company. Control are sensitive to that, and it will just be old friends who visit. They more than anyone else know what we've gone through.

I drifted into the cupola last night and it was totally dark inside. Staring at Earth, I could see the whole blue planet reflected in my eye and my eye reflected in the window. It was perhaps the most beautiful sight I'd seen in six years.

✳ Tom Kirby, Commander
2242 days, 20 hours, 21 minutes

Last night I remembered the promises we made that snowy January day more than six years ago, when we pledged to carry the human spirit of exploration to the planets. It was a promise none of us were completely sure we could fulfil. There was so much that could go wrong, that could snuff out this daring attempt to stray beyond Earth. And yet, somehow, we reached for the planets and returned to tell the tale.

Project *Pegasus* is over, but I hope that our exploration of the planets will continue. The journey into space must go on, and I pray that one day I will return to that place where the blue sky turns black.

The crew in Earth orbit once more take in the view of their home planet as if it were another alien world.

Earth

The third rock from the Sun sits comfortably in the solar system's so-called 'Goldilocks zone', where solar temperatures make it habitable. The atmosphere, made mostly of nitrogen and oxygen, shields the surface from radiation and small meteorites, and maintains the conditions for water to exist in all three of its phases: ice, liquid and vapour. More than 70 per cent of Earth's surface is covered by liquid water, which underpins all the geological and chemical cycles that permit life to thrive. Tilted at 23.5 degrees, Earth alternately presents its northern and southern hemispheres to the Sun, creating the seasons. Pulled by the Moon and the Sun, the axis of rotation wobbles in a circle, taking 25,000 years to rotate once. Earth's crust is broken into plates that form the continents and ocean basins. These plates move over the surface in a process called plate tectonics, driven by convection currents rising through a hot mantle of semimolten rock beneath. The planet's iron core is also molten and its movement generates a magnetic field strong enough to protect the surface from harmful solar radiation. Earth can be considered as a double planet system, with one large moon a third as wide as itself.

1. A hurricane brews beneath *Space Shuttle Endeavour* in September 1995.

2. An image of Earth taken by *Endeavour* in 1993.

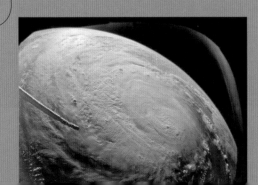

So *Pegasus* returned with no convincing evidence of life from the other planets in our solar system. I don't think many of us were that surprised. Evidence for extraterrestrial life has always proved hard to confirm. I still believe that there is every chance that our planetary system once had multiple niches for life, but it was always going to be a long shot to find anything still living on these poisonous, frozen worlds that are our neighbours. Maybe one planet stuffed full of organic life forms might be enough for any planetary system.

So do I believe we are alone in the Universe? Certainly not, as astronomer Carl Sagan was fond of saying 'absence of evidence is never evidence of absence'. We might have trawled our own planets and moons for signs of life, but the real search for extraterrestrial life lies across the Galaxy in other planetary systems. And with the Lowell telescope now in place on Pluto we have an extraordinary opportunity to study these new worlds in orbit around their own stars.

Even before *Pegasus* left Pluto's orbit the Lowell telescope began sending back images of 'extrasolar' planets. New pictures poured in each week. They only appeared as unnamed specks of light, but I found them tantalizing nonetheless; they are worlds out of our reach. Their light has taken at least a human lifetime, sometimes much more, to reach our telescope. Unfortunately we will never reach such far-flung worlds in my children's lifetime, let alone my own. But I feel content that with *Pegasus* we have begun this important new search for life amongst the stars.

Perhaps, one day, when advances in human ingenuity and technology allow, a host of spacecraft – the successors to *Pegasus* – will carry our descendants to the stars. And if they don't find life in one solar system, it won't stop them looking in others. Maybe these future pioneers of interstellar travel will even remember us – the first humans to step onto the planets and the generation who left them this legacy of new worlds to explore.

Alex Lloyd
Chief Scientist, *Pegasus* Mission

The Star 51-Pegasi in the constellation of Pegasus is 50 light years away from us. It was the first star like our Sun seen to be wobbling because of a planet or planets orbiting it. The major planet is half as big as Jupiter and takes just over four days to orbit the star. This first extrasolar planet to be detected kicked off a Copernican revolution in astronomy. Astronomers have now discovered more than a hundred extrasolar planetary systems using this technique. The interplanetary spacecraft in our story, *Pegasus*, was named in honour of this great planetary discovery.

ACKNOWLEDGEMENTS

Although there are plenty of words in this book, it is the pictures that bring this vision of a human future in space alive. Without the talent and perseverance of an exceptional team of artists led by Mike Milne at Framestore CFC in London, the story we attempt to tell here would be a far less vivid version of the one bound up in this book you are holding now. Among the huge team of talented artists we would particularly like to thank: the brilliant spacecraft modellers Sarah Tosh, Jon Veal, Romain Segurado and Oliver Cook; the very gifted digital paint artists Daren Horley, Jason Horley and Nathan Hughes; and the TDs Angela King and Henriette Plum.

We consulted many scientific advisors in researching the story. Dr Richard Taylor of the British Interplanetary Society was the first to share his concepts of such a human mission with us very early on. Space-flight historian and former spacecraft engineer Dr David Baker helped us with the initial designs for *Pegasus* and our landers and the drafting team at EADS Space generously took time out from creating real spacecraft to help us perfect our interplanetary leviathan.

We also owe a lot to astronaut-explorer Dave Scott who commanded the *Apollo 15* mission to the Moon. During the making of the series he shared with us his own historic experiences of standing on another world and staring back at Earth, and brought a reality to this story that few other human beings could have provided.

From very early on the European Space Agency has been an extremely supportive partner, ultimately lending us their human space flight mission control complex in the Netherlands for a week to film our own mission control scenes. In particular we are indebted to Claus Habfast, Dieter Isakeit, Massimo Sabbatini, Marie-Noelle Oostijen, Muzzammal Raouf and Zoe Williamson. At NASA Bobbie Faye-Ferguson ensured we had access to scientists, engineers and astronauts who could point us in the right direction.

Although much of the series was digitally generated, there were a lot of props that were built from scratch and we'd like to thank Jamie Campbell and his team for the often back-breaking work they put in to help our dreams materialize. And without the ceaseless work and thoughtful support of Sam Jukes-Adams, Sarah Baxter and Karina Randall, masterminding the mind-boggling logistics needed to haul a sizeable team of people and props across several continents, we would have fallen at the first hurdle.

In Chile, where we created our Mars and Venus scenes, we are indebted to Arturo Opaso and Jonathan Franklin and their tireless team of drivers and fixers. And in Russia – where we lived the miracle of real weightlessness and took over the Yuri Gagarin Cosmonaut training facilities at Star City – we could not have accomplished a thing without Marina Erastova and Lieutenant Sergey Prokopovich.

Finally - without the patience of five actors who played our main cast this story would not have come to life. We wish to thank Martin McDougall, Rad Lazar, Joanne Macinness, Michelle Joseph and Mark Dexter for enduring the hardships of uncomfortable space suits and extreme environments inflicted on them in the filming of our odyssey. Their characters and performances were carefully nurtured by a very dedicated director, Joe Ahearne.

Any project like this takes over your life for months on end and without the patience and understanding of friends and families our lives might now be without either.

Chris would like to thank his treasured wife Jacqui Farnham and his dear friends Duncan Copp and Rajeev Thacker for their unstinting support and encouragement throughout, and for still being there when he came back from the mission.

Tim would like to thank his dear wife Clare and his four little ones Olivia, Eleanor, Angus and Rufus. In addition he would like say he is very much indebted to Chris Riley, whose enthusiasm, energy and expertise has underpinned this entire project.

INDEX

BBC Books would like to thank the following for providing photographs and for permission to reproduce copyright material. While every effort has been made to trace and acknowledge all copyright holders, we would like to apologize for any errors or omissions.

BBC 2, 4–5, 10–11,12, 13 above, 13 below, 14 above, 14 below, 15, 17, 18–19, 22, 26 above, 26 below, 26–7, 28, 30, 31, 34, 38, 39, 40, 41, 46, 51, 54–5, 58, 59, 60, 61, 62 above, 62 below, 66 left, 66 right, 67, 70–1, 73, 74 left, 75, 78–9, 82, 83, 86, 88, 92–3, 100, 101 above, 101 below, 104 left, 104 right, 107, 108, 109, 110–11, 112, 113 below, 114–15, 118–19, 120, 121, 124, 125, 127, 131, 132, 134, 135 left, 136 below, 137, 140–1, 142–3, 144, 145, 146, 147, 156–7, 158, 159, 169, 171, 173, 174, 175, 178, 179, 182, 183; **The Bridgeman Art Library** 24, 76 above, 76 below; **Don Campbell /Aricibo** 36 top; **Corbis** 77 below, 180 above, 184–5; **ESA** 69 below right, 90 left; **Galaxy Picture Library** 68 below, 69 above centre, 69 above right, 81 above right, 98, 129 right, 138 above, 148 centre, 155 below left, 155 centre right, 161 below left, 161 right, 166 below, 167 above, 167 centre, 167 below, 176 left, 176 right, 177 left, 177 right; **Jodrell Bank Observatory** 25; **Lowell Observatory** 149 below; **NASA** 7 above centre, 7 above centre right, 7 below left, 7 below centre right, 9, 20–1, 33 below, 36 right, 36 centre, 36 below, 37, 42 left, 43 below left, 43 below right, 52 above, 52 below, 53 left, 53 right, 56 above left, 56 above right, 56 centre, 56 below, 57 above right, 57 below, 63, 64–5, 68 above, 69 above left, 69 below left, 74 right, 81 left, 81 centre right, 81 below, 84 above, 85 left, 85 centre, 85 below, 91 above left, 91 above right, 91 centre, 91 below, 94, 95 left, 95 right, 96 above, 96 below, 97, 102 left, 103 centre, 103 centre right, 103 below right, 105 above, 105 below, 113 above, 116 above left, 116 centre left, 116 centre right, 116 below right, 117 left, 117 centre, 117 right, 123 left, 123 below right, 128 left, 128 right, 129 left, 136 above, 138 centre left, 138 centre right, 138 below, 139 left, 139 right, 155 above right, 160, 161 above left, 181 above left, 181 above right, 181 below left, 181 below right, 184 left; **NASA/ Don P. Mitchell** 33 left, 33 centre, 33 right, 42 above, 42 right, 43 right; **Chris Riley** 149 left, 149 right; **Royal Astronomical Society** 48 left, 77 above, 99, 122 right, 166 above; **Royal Observatory Edinburgh/Anglo-Australian Observatory** 187; **Science Photo Library** 7 above left, 7 above centre left, 7 above right, 7 below centre left, 7 below right, 24–5, 32 right, 44–5, 48 right, 80, 84 below, 89, 90 right, 122 left, 122 below, 123 above right, 139 centre, 148 left, 155 below right, 160 right, 162–3, 164–5, 177 centre, 180 below, 184–5; **STSCI/NASA** 32 left, 102 above centre, 102 below centre, 102 below, 154 left, 154 right, 155 above left; **United States Airforce Chart and Information Centre** 49